AF598104

MICRO/ NANOLITHOGRAPHY - A HEURISTIC ASPECT ON THE ENDURING TECHNOLOGY

Edited by **Jagannathan Thirumalai**

Micro/Nanolithography - A Heuristic Aspect on the Enduring Technology
http://dx.doi.org/10.5772/intechopen.68234
Edited by Jagannathan Thirumalai

Contributors

Amalraj Peter Amalathas, Maan Alkaisi, Jagannathan Thirumalai, Yanqing Wu, Shumin Yang, Hongbo Lan, Sujatha Lakshminarayanan, Ahmed Nassouh Awad, Atushi Takahashi, Chikaaki Kodama

First published in London, United Kingdom, 2018 by IntechOpen
IntechOpen is the global imprint of INTECHOPEN LIMITED, registered in England and Wales, registration number: 11086078, The Shard, 25th floor, 32 London Bridge Street
London, SE19SG – United Kingdom
Printed in Croatia

British Library Cataloguing-in-Publication Data
A catalogue record for this book is available from the British Library

Additional hard copies can be obtained from orders@intechopen.com

Micro/Nanolithography - A Heuristic Aspect on the Enduring Technology, Edited by Jagannathan Thirumalai
p. cm.
Print ISBN 978-1-78923-030-7
Online ISBN 978-1-78923-031-4

For EU product safety concerns:
IN TECH d.o.o., Prolaz Marije Krucifikse Kozulić 3, 51000 Rijeka, Croatia,
info@intechopen.com or visit our website at intechopen.com.

Meet the editor

Dr. J. Thirumalai received his PhD degree from Alagappa University, Karaikudi, in 2010. He was also awarded with the postdoctoral fellowship from Pohang University of Science and Technology (POSTECH), Republic of Korea, in 2013. He worked as an assistant professor in Physics, BS Abdur Rahman University, Chennai, India (2011 to 2016). Currently, he is working as an assistant professor of Physics, SASTRA Deemed University, Kumbakonam (TN), India. His research interests focus on luminescence, self-assembled nanomaterials and thin-film optoelectronic devices. He has published more than 50 Scopus/ISI-indexed papers and 10 book chapters and edited 4 books. He is a member of several national and international societies like RSC, OSA, etc. He has edited two books for IntechOpen publisher. He served as the principal investigator for a funded project towards the application of luminescence-based thin-film optoelectronic devices, funded by the Science and Engineering Research Board (SERB), India.

Contents

Preface

Fabrication of micro- and nanostructure has opened new horizons in science and engineering by employing novel lithographic techniques. The success of developing and manufacturing integrated circuits and micro- and nanodevice prolonged the innovative technological advancements in nano- and microtechnologies. Nanofabrication is a 'gating' technology for the achievement of all future advanced nanodevices. Lithography is more fascinating and important to mankind since the prehistoric times are named after the art of printing from stone to new material. Capitalizing on the latest origination of novel applications of lithography has proffered innovative technological advancements, which are rapidly implemented for ground-breaking research, especially in the sub-nanoscale, which studies the physical and chemical properties of lithographic process.

Several optoelectronic micro- and nanodevices and their systems employed in contemporary industry are decreasing in size and have reached the nanoscale range. Micro-/nanofabrication is critical to the consciousness of the prospective benefits of innovative devices and their systems for social order. The evolution of inventive lithographic technologies almost rests on the utilization of the current physico-chemical materials. The basis of contemporary industrial lithographic technologies that generate employments and superior track record will be extended with new innovations by showing a significant improvement in comparison with multiple patterning—extreme ultraviolet (EUV) and directed self-assembly (DSA). In the structural artifact, the process of learning with respect to lithographic techniques would lead to innovative applications to the scientific community, which has the prospect to improve the quality of life.

In the present day, the latest lithographic methods provide an energetic, developing scientific discipline at the vanguard of thin-film material sciences with good international scope and connections.

The nanolithography technique is used for patterning in the nanoscale region, and this technique comprises photolithography, electron beam lithography (EBL), X-ray lithography, extreme ultraviolet lithography (EUVL), light coupling nanolithography (LCM), scanning probe microscope lithography (SPM), nanoimprint lithography, dip-pen nanolithography and so on. Besides semiconductors, non-semiconductor nano- and microtechnologies, e.g. MEMS, sensors and magnetic storage media, are emerging and will eventually find their place in markets as well. Furthermore, lithographic techniques find a wide range of applications in everyday life, starting from modern fluorescent lighting to light emitting diodes (LEDs), solar concentrators and many other electrical and electronic equipments. The lithography necessities for these technologies are so often entirely dissimilar from the semiconductor architectures. As a result, the aforementioned additional techniques of lithography may perhaps be employed for these applications. The capability to imitate lithographic patterns from microscale to nanoscale is of pivotal prominence to the advancement of micro-/nanoscience and its technologies. This book would form the basis for a better fundamental understanding of the capabilities and limitations of each type of lithography and may also suggest better, cheaper or alternative lithography technologies to be considered in different applications.

The objective of this book is to offer timely and in-depth coverage of designated advanced topics in lithographic techniques by eminent contributors. Also, researchers from various fields are often working in the field of lithography. This book drives into structural components concerning the many elements of knowledge necessary to comprehend both quantitatively and qualitatively the cutting-edge subjects in lithographic methods.

The chapters for this book have been contributed by esteemed researchers in fields of lithographic fabrication and technology and cover the frontier areas of research and developments in the field of micro- and nanolithographic science and technology. This book is focussed at students and researchers wishing to gain a deeper understanding about the many fundamental concepts that are of concern in relevant fields, and it is our anticipation that you, the readers, will find this book useful for your work.

This book consists of six chapters covering topics related to the eminence of lithography, the types of lithography with necessary examples, fabrication and reproduction of periodic nanopyramid structures using UV nanoimprint lithography for solar cell applications, large-area nanoimprint lithography and applications, micro-/nanopatterning on polymers, OPC under immersion lithography associated to novel luminescence applications and EUV/soft X-ray interference lithography.

I would like to thank all the contributors in this book for their tremendous efforts in delivering an outstanding work. Last but not least, I would like to express my sincere gratitude to Ms. Romina Rovan, IntechOpen publishing process manager, for the effective communication and assistance during the preparation of this book.

Jagannathan Thirumalai
SASTRA Deemed University, India

Chapter 1

Introductory Chapter: The Eminence of Lithography—New Horizons of Next-Generation Lithography

Jagannathan Thirumalai

Additional information is available at the end of the chapter

http://dx.doi.org/10.5772/intechopen.70725

1. Introduction

1.1. A laconic antiquity on lithography

Over the last three centuries the term "lithography" (from the ancient Greek *lithos*, meaning "stone," and *graphein*, meaning "to write") has been adopted [1]. And photogravure is a process that uses a stone (in general lithographic limestone) or the smooth surface of a metal plate. The printing technique of lithography was first invented by the German playwright and actor Alois Senefelder in the Kingdom of Bavaria in 1796, and was a viable method for publishing histrionic works [2, 3]. Lithography could be used to pattern a script or artwork on paper or other suitable material [4]. Only the stone parts would absorb the liquid; the design parts repelled it. Rolling on ink consisting of soap, wax, oil, and lampblack, the greasy material, which was coated over the pattern, could not cover the surface that was repelled by moisture in the blank areas. As soon as a sheet of paper was applied over the surface of the stone, a clean impression of the design was produced. Lithography established its popularity throughout the mid-1900s because the process inspired printers to discover additional practical and quicker techniques of printing drawings [5]. The history of lithography came about in four major steps: (1) the invention and early usage of the process; (2) the introduction of photography related to the process; (3) the addition of the offset press corresponding to the process; and (4) the discovery of the lithographic plate [6].

In 1850, the first *steam litho* press was invented by R. Hoe in France and was popularised in the United States in 1868 [7]. Lithographic stones were used to prepare the image and a cylinder covered with a blanket received the image from the plate, which was transformed to the respective substrate. Direct rotary presses used for lithography were comprised of zinc and aluminium metal plates, which were first produced in the 1890s. The first offset press was developed during 1906 by Ira W. Rubel [8] (who was a paper maker). From a press cylinder, an imprint was

S. No.	Techniques	Patterning methods	Optimum environments	Resolution	Merits	Limits	Examples
1.	Microlithography and nanolithography	Creates patterns by structuring material on a fine scale	Vacuum	10 μm and 100 nm	• Efficient and cost effective	• Several processing steps • More complexity	Double/multiple patterning lithography [10, 11]
2.	Contact lithography (CL)	Image printed is obtained by illumination of a photomask in direct contact with a substrate coated with an imaging photoresist layer	Vacuum	~ 100–1000 nm	• Lower-cost process • Stress-free usage	• Oxidation of the metal surface destroys plasmon resonance conditions	Fabrication of metal ring arrays on silicon substrate [12]
3.	Scanning probe microscope (SPM) lithography	A direct-write, maskless approach that bypasses the diffraction limit based on tip–sample interaction	Ambient vacuum or liquid phase	Below 50 nm	• Cost effective • Stress-free usage • Suitable for a wide range of materials • High sensitivity and efficiency	• Controllability and accessibility for large-scale production • Serial patterning	Micrometer-scale SPM local oxidation using the micrometer tip under contact-mode operation [13]
4.	Optical photolithography (OPL)	A lithographic printing process that selectively exposes plates or substrate to UV radiation for the formation of images	Vacuum	Usually at sub-100 nm[4]	• Highly efficient and cost effective • Controls the exact size and shape of the entire substrate • Pattern is parallel in nature	• High operation cost • Multiple processing steps	Structures with silver film were used as the exposure mask [14]
5.	Electron beam lithography (EBL)	Direct writing of structures down to sub-10 nm dimensions, and also facilitating high-volume nanoscale patterning technologies	Vacuum	High resolution up to sub-10 nm (maximum of ≤50 nm)	• Prints complex patterns directly on wafers • Eliminates the diffraction problem • Flexible technique	• Slower than optical lithography • Expensive and complicated	Recent developments in processing, tooling, resist and pattern [15]

S. No.	Techniques	Patterning methods	Optimum environments	Resolution	Merits	Limits	Examples
6.	Focused ion beam lithography (FIBL)	Consists of a focused beam of ions that can be operated at low beam currents for imaging or at high beam currents for site-specific sputtering	Vacuum	≤50 nm	• High sensitivity and efficiency • Diffraction effects are minimised • Less backscattering	• High operation cost • Multiple processing steps • Poor accessibility	Structuring approaches of novel patterns [16]
7.	Extreme ultraviolet lithography (EUVL)	Consists of burning intense beams of ultraviolet light that are reflected from a circuit design (semiconductor integrated circuits (ICs)) pattern into a wafer	High vacuum	≤13.5 nm	• Excellent multipatterning and additional layers • Eliminates the diffraction problem • With low cycle times enhances yields on preparing IC chips • Helps produce smaller feature size	• Increased cost for new technology • Several processing steps • Complex	Next-generation semiconductor [17]
8.	Light coupling mask nanolithography (LCML)	Consists of a polymer mask placed in contact with the photoresist through transparent regions that protrude through the topographically patterned mask where exposure is required for obtaining the image	Vacuum	≤50–20 nm (365 and 436 nm)	• Lower-cost process • High density High optical resolution	• Wider gap between the mask and the substrate can cause images based on evanescent waves • Oxidation of the metal surface also extinguishes plasmon resonance	Organic polymers assist amplitude mask for light-based lithographies [18]
9.	X-ray lithography (XL)	Uses X-rays to transfer a geometric pattern from a mask to a light-sensitive chemical photoresist on the substrate	Vacuum	≤20 nm	• Large-area patterning across an A4-size area • Not affected by organic defects in mask • Reduction in diffraction, reflection and scattering effects • Shorter wavelengths (0.1–10 nm)	• Shadow printing • Lateral magnification error • Brighter X-ray sources needed • More sensitive resists needed	Lithographic beam lines for soft and hard X-ray for micro- and nanofabrication [19]

S. No.	Techniques	Patterning methods	Optimum environments	Resolution	Merits	Limits	Examples
10.	Nanoimprint lithography (NIL)	Creates patterns by mechanical deformation of imprint resist and subsequent processes	High vacuum or ambient	Resolution up to ~100 nm	• Low cost • High throughput • High resolution	• Precession issue • Multiple steps for large-scale production	Polymer material (h-PDMS) [20]
11.	Dip-pen nanolithography (DPN)	Direct-write patterning technique based on atomic force microscopy (AFM) scanning probe technology on a range of substances with a variety of inks	Vacuum	Minimum resolution up to ~50 nm	• High density • Lower cost • High-throughput printing through organic and inorganic inks	• Smooth surfaces to work on • Write head can be turned on/off at will	Molecular electronics to materials assembly [12]
12.	Neutral atomic beam lithography	Creates patterns by using a neutral atomic beam to create permanent structures on surfaces	Vacuum	~70 nm	• High-throughput printing • Novel lithographic schemes based on the optical quenching of internal energy	• Multiple steps for large-scale production	Self-assembled monolayers of alkanethiolates on Au and alkylsiloxanes on SiO_2 [21]
13.	Interference lithography	Creates patterns by regular arrays of fine features, without the use of complex optical systems or photomasks	Vacuum	~50 nm	• Quick generation of dense features over a wide area without loss of focus	• Limited to uniformly distributed aperiodic patterns only • Affected by electron interference lithography non-optical effects	3D photonic crystals [22]
14.	Hot-embossing lithography	Creates patterns using polymer or glass substrates to imprint structures created on a master stamp	Vacuum	~50–100 nm	• Low cost • Flexible fabrication • High resolution	• Controllability and accessibility for large-scale production	Polymer-based interdigitated electrodes [23]

S. No.	Techniques	Patterning methods	Optimum environments	Resolution	Merits	Limits	Examples
15.	• Projection microstereo-lithography (PμSL) • (3D printing technology)	Creates patterns using rapid photopolymerisation of an entire layer with a flash of UV illumination at microscale resolution; in addition, the mask can control individual pixel light intensity, allowing control of material properties of the fabricated structure with desired spatial distribution	Ambient temperature and atmosphere	~500 μm	• Enables integration of multiple material elements in a single process	• High operation cost • Multiple steps for large-scale production	Lincoln Monument [24]
16.	Charged-particle lithography	Used for creating patterns; the imaging action is mediated by charged particles such as electrons (as in EBL) and ions (as in ion beam lithography).	High vacuum or ambient	~50–100 nm	• High sensitivity and efficiency • Diffraction effects are minimised • Flexible fabrication • High resolution	• Slower than optical lithography • Expensive and complicated	Fabrication of electronic devices and microstructures using high-resolution organic resists [25]
17.	Neutral-particle lithography	Used for creating patterns; a broad beam of energetic neutral atoms floods a stencil mask and transmitted beamlets transfer the mask pattern to resist on a substrate	Vacuum	~50–100 nm	• High resolution • High-throughput printing • Novel lithographic schemes based on the optical quenching of internal energy	• Multiple steps for large-scale production	Bird's-eye view of a 50 nm wide slot [26]
18.	Atomic force microscopic nanolithography or scanning force microscopy (SFM)	The simplest way to attain single structure formation in which the tip is immobilised at a specific surface site, and a large force is then applied to the tip to indent the surface	Ultra-high vacuum (UHV)	~100 nm	• For visualising samples that do not require any special treatments such as metal/carbon coatings • AFM can provide higher resolution than SFM	• Height of 10–20 μm (resolution) • AFM probes cannot normally measure steep walls or overhangs	Development of more complex nanodevices such as single-electron transistors [27]

S. No.	Techniques	Patterning methods	Optimum environments	Resolution	Merits	Limits	Examples
19.	Magneto-lithography	Creates patterns based on the magnetic field on a substrate, using paramagnetic or diamagnetic masks, that defines the shape and strength of the magnetic field	High vacuum or ambient	~100 nm	• Simple • High resolution • High-density patterned surfaces	• Less expensive • Multiple steps for large-scale production	Magnetic Fe_3O_4 nanoparticles pattern on a gold thin film [28]
20.	Multibeam or complementary E-beam lithography (CEBL)	Uses multiple miniature columns and vector scanning of shaped beams (critical layers) to boost throughput	UHV	~50–100 nm	• High resolution • High-throughput printing • Novel lithographic schemes based on the optical quenching of internal energy	• Controllability and accessibility for large-scale production	Development of more complex nanodevices [29]
21.	Scattering with angular limitation in projection electron beam lithography (SCALPEL)	Creates patterns with extremely small features in microelectronic circuits. Electrons are projected onto a "mask", which then pass straight through the mask, transferring the image of the mask to the wafer	UHV	~70 nm	• High resolution • Novel lithographic schemes based on the optical quenching of internal energy • High-quality patterned images	• High operation cost • Multiple steps for large-scale production • Controllability and accessibility for large-scale production	For semiconductor manufacturing lithography with feature sizes beyond the capabilities of optical lithography [30]

Table 1. List of various lithography techniques in the nanometer and micrometer range.

http://dx.doi.org/10.5772/intechopen.70725

inadvertently printed over the impression cylinder's rubber blanket. Once a sheet of paper was run along the press, an intense image was printed on it using the imprint that was being counterpoised on the rubber blanket. A.F. Harris, the inventor of offset lithography, noticed a similar effect. He then established an offset press applicable for the Harris Automatic Press Company in the same year. Harrold and Wright [9] invented the offset process and created the most familiar method of offset lithography from the 1925s to the 1950s using enhanced plates, inks (multicolour), multicylinders, papers, etc. In the late 1950s, offset lithographic printing dominated all other offset printing methods because it produced sharper, clearer images than letterpress and also cost less when compared to engraving. Currently, the mainstream of offset lithographic printing (more than 50%), including newspapers, is mainly produced by using offset printing methods. Lithography, as well as the planographic printing method, makes the best use of the incompatibility of water and grease. In the offset lithographic technique, liquid/powder ink is coated onto a grease-treated image over the flat printing surface; the blank portions that attract moisture repel the lithographic ink. **Table 1** summarises the various lithography techniques in the nanometer and micrometer ranges and the prediction of innovative occurrences [10–31].

2. Next-generation lithography in the new skylines of science and engineering

Fabrication on micro- and nanoarchitectures has opened new horizons in the area of engineering, science and technology. The success of improving and yielding micro- and nanodevices and integrated circuits (ICs) by using photolithography practices is prominently incredible. Nanofabrication is considered as a "gating" technology for the accomplishment of all future advanced nanodevices. Over the past two decades, photolithography had been broadly used for the purpose of microdevices and integrated circuit (IC) technology. However, the wavelength of photons and the search for optics and resistance offered by the materials have limited the resolution of nanostructures prepared from lithography to about 100 nm. In addition, next-generation lithography (NGL) processes, such as maskless, E-beam and direct write lithography, require specific and expert intervention to open up new product/market combinations. The movement towards 450 mm wafers presents its own set of challenges. The larger wafers require new processes, and equipment cost control is a key concern. The equipment needed to support these techniques needs to be as precise and reliable as the chips they make. There are five important candidates for NGL [32] technology to ensure rigorous growth: (1) X-ray proximity, (2) extreme ultraviolet lithography (EUV), (3) ion projection lithography (IPL), (4) scattering with angular limitation projection electron-beam lithography (SCALPEL), and (5) nanoimprint lithography (NIL). NGL technology would be familiarised and alike adept for the probable imminent in a mix-and-match mode along with optical lithography. Continuing developments based on NIL are leading semiconductor manufacturers to use this technological development as a probable auxiliary for optical lithography, but which may be limiting due to its reduced capacity [33, 34]. As a consequence, the existing tumult in the fabrication of novel technological developments is associated with the development of innovative thoughts on the emerging field of nanotechnology, high-power LEDs, nano/micro-ICs and so on.

A maskless NGL tool could meet the following requirements: cycle time, mask cost, removal of the mask input to the compact disk (CD) control, etc. With a comparable industrial price for numerous wafers per mask, there would be a need for this NGL implement by means of no issue whether the manufactured goods is logic, flash, or Dynamic Random Access Memory (DRAM). This trend continues uninterrupted. As followed by Moore's law, it is required to follow the incessant progressions in lithographic steadfastness.

3. Conclusion

In light of the aforementioned discussion, lithography has shown that, with high novel scientific achievements and research, it may outshine other innovative applications for the purpose of common interest. Among the major notable growth areas are the diverse fields of nanotechnology, photovoltaics/solar cells, displays, LEDs and eco-friendly materials. Differences in innovative lithographic printing techniques continue to be improved through novel global progressions in spectral imaging, time-correlated single-photon counting, noninvasive optical biopsy, visual implants and kinetic chemical reaction rates. Hence, research on exclusive lithographic printing technological accomplishments would lead the way to more effective techniques, while coating thin films at the atomic level may turn out to be the ideal printing technology for the future.

Acknowledgements

The authors apologise for inadvertent omission of any pertinent references.

Conflict of interest

The authors declare that there is no conflict of interest related to the publication of this chapter.

Author details

Jagannathan Thirumalai

Address all correspondence to: thirumalaijg@gmail.com

Department of Physics, School of Electrical and Electronics Engineering, SASTRA Deemed University, Srinivasa Ramanujan Center, Kumbakonam, India

References

[1] Peter Weaver. The Technique of Lithography. 1st ed. Michigan: London: Batsford; 1964. pp. 176 p

[2] Philip B. Meggs. A History of Graphic Design. 2nd ed. USA: International Thomson Publishing Company; 1992. pp. 528 p. ISBN: 0442318952

[3] Rob Carter, Philip B. Meggs, Ben Day, Sandra Maxa, Mark Sanders. Typographic Design: Form and Communication. 6th ed. Germany: John Wiley & Sons, Inc.; 2014. pp. 352 p. ISBN: 978-1-118-71576-5

[4] Elizabeth Robins Pennell. Lithography and Lithographers. 1st ed. New York: The Macmillan Company: London; 1915. pp. 366 p. DOI:

[5] Printing History – History of Lithography [Internet]. April 29th, 2016 [Updated: April 29th, 2016]. Available from: https://dynodan.com/magazine-printing/process/printing-history/printing-history-history-of-lithography/ [Accessed: 24-02-2017]

[6] The History of Lithography [Internet]. Available from: http://sites.tech.uh.edu/digitalmedia/materials/3350/History_of_Litho.pdf

[7] Stephen D. Tucker. History of R. Hoe & Company, 1834-1885. In: Rollo G. Silver, editor. Proceedings of the American Antiquarian Society; October 1972; Worcester, Massachusetts, USA: http://www.americanantiquarian.org/proceedings/44498049.pdf; 1973. p. 349-453. DOI: NA

[8] Ira Rubel Invents the First Offset Press [Internet]. 2004-2017. Jeremy Norman & Co., Inc. [Updated: June 16, 2017]. Available from: http://www.historyofinformation.com/expanded.php?id=666 [Accessed: July 3, 2017]

[9] Harrold CW, Wright CJ "Printing Press". U.S. Patent 1,591,127 filed April 4, 1922 and issued July 6, 1926

[10] Bruce W. Smith Kazuaki Suzuki, editors. Microlithography: Science and Technology. 2nd ed. Boca Raton, Florida, US: CRC Press; May 11, 2007. pp. 541 p. DOI: 10.1201/9781420051537

[11] David Z. Pan, Bei Yu, Jhih-Rong Gao. Design for manufacturing with emerging nanolithography. IEEE Trans. Comput.-Aided Design Integr. Circuits Syst. Oct. 2013;**32**(10): 1453-1472. DOI: 10.1109/TCAD.2013.2276751

[12] Woongkyu Park, Jiyeah Rhie, Na Yeon Kim, Seunghun Hong, Dai-Sik Kima. Sub-10 nm feature chromium photomasks for contact lithography patterning of square metal ring arrays. Sci. Rep. 2016;**6**(1):23823. DOI: 10.1038/srep23823

[13] Jun-ichi Shirakashi. Scanning probe microscope lithography at the micro- and nanoscales. J. Nanosci. Nanotech. 2010;**10**(7):4486-4494. DOI: 10.1166/jnn.2010.2359

[14] Li Hai-Hua, Chen Jian, and Wang Qing-Kang. Research of photolithography technology based on surface plasmon. Chin. Phys. B. 2010;**19**(11):114203-114205. DOI: 10.1088/1674-1056/19/11/114203

[15] Ampere A. Tseng, Kuan Chen, Chii D. Chen, Kung J. Ma. Electron beam lithography in nanoscale fabrication: recent development. IEEE Trans. Electron. Packag. Manuf. 2003; **26**(02):141-149. DOI: 10.1109/TEPM.2003.817714

[16] Heinz D, Wanzenboeck, Simon Waid. Focused ion beam lithography. In: Bo Cui, editor. Recent Advances in Nanofabrication Techniques and Applications. 2nd ed. Croatia, Europe: InTech Open Access; 2011. p. 1-614. DOI: 10.5772/22075

[17] Toshiro Itani, Takahiro Kozawa. Resist materials and processes for extreme ultraviolet lithography. Jpn. J. Appl. Phys. 2013;**52**(1):010002-01000214. DOI: 10.7567/JJAP.52.010002

[18] Heinz Schmid, Hans Biebuyck, and Bruno Michel. Light-coupling masks for lensless, sub-wavelength optical lithography. Appl. Phys. Lett. 1998;**72**(1):2379. DOI: 10.1063/1.121362

[19] Di Fabrizio E, Fillipo R, Cabrini S, Kumar R, Perennes F, Altissimo M, Businaro L, Cojac D, Vaccari L, Prasciolu M, Candeloro P. X-ray lithography for micro- and nano-fabrication at ELETTRA for interdisciplinary applications. J. Phys.: Condens. Matter. 2004;**16**(33):S3517. DOI: 10.1088/0953-8984/16/33/013

[20] Weimin Zhou, Guoquan Min, Jing Zhang, Yanbo Liu, Jinhe Wang, Yanping Zhang, Feng Sun. Nanoimprint lithography: a processing technique for nanofabrication advancement. Nano-Micro Letters. 2011;**3**(2):135-140. DOI: 10.3786/nml.v3i2

[21] Khalid Salaita, Yuhuang Wang, Chad A. Mirkin. Applications of dip-pen nanolithography. Nat. Nanotechnol. 2007;**2**(1):145-155. DOI: 10.1038/nnano.2007.39

[22] Thywissen JH, Johnson KS, Younkin R, Dekker NH, Berggren KK, Chu AP, Prentiss M. Nanofabrication using neutral atomic beams. J. Vac. Sci. Technol. B. 1997;**15**(6):2093-2100. DOI: 10.1116/1.589227

[23] Andreas J. Wolf, Hubert Hauser, Volker Kübler, Christian Walk, Oliver Höhn, and Benedikt Bläsi. Origination of nano- and microstructures on large areas by interference lithography. Microelectron. Eng. 2012;**98**(October):293-296. DOI: 10.1016/j.mee.2012.05.018

[24] Schift H, Jaszewski RW, David C, Gobrecht J. Nanostructuring of polymers and fabrication of interdigitated electrodes by hot embossing lithography. Microelectron. Eng. 1999;**46**(1):121-124. DOI: 10.1016/S0167-9317(99)00030-1

[25] Muskin J, Ragusa MJ. Three-dimensional printing using a photoinitiated polymer. Chem. Educ. 2010;**87**(5):512-514. DOI: 10.1021/ed800170t

[26] Yukinori Ochiai, Shoko Manako, Jun-ichi Fujita, Eiichi Nomura. High resolution organic resists for charged particle lithography. J. Vac. Sci. Technol. B: Nanotechnol. Microelectron. 1999;**17**(3):933. DOI: 10.1116/1.590672

[27] Wolfe JC, Craver BP. Neutral particle lithography: a simple solution to charge-related artefacts in ion beam proximity printing. J. Phys. D: Appl. Phys. 2008;**41**:024007-024012. DOI: 10.1088/0022-3727/41/2/024007

[28] Notargiacomo A, Foglietti V, Cianci E, Capellini G, Adami M, Faraci P, Evangelisti F, Nicolini C. Atomic force microscopy lithography as a nanodevice development technique. Nanotechnology. 1999;**10**(4):458-463. DOI: 10.1088/0957-4484/10/4/317

[29] Bardea A, Yoffe A. Magneto-lithography, a simple and inexpensive method for high-throughput, surface patterning. IEEE Trans. Nanotechnol. 2017;**16**(3):439-444. DOI: 10.1109/TNANO.2017.2672925

[30] Multibeam News. Multibeam: Applications – Industry Overview [Internet]. July 2017 [Updated: July 2017]. Available from: http://www.multibeamcorp.com/applications.htm [Accessed: July 7, 2017]

[31] Lloyd R. Harriott. Scattering with angular limitation projection electron beam lithography for suboptical lithography. J. Vacuum Sci. Tech. B: Nanotech. and Nanometer Structures. 1997;**15**(6):2130-2135. DOI: 10.1116/1.589339

[32] EDITORIAL: "Next-generation lithography", J. Micro/Nanolith. MEMS MOEMS. **6**(4): 040101 (December 27, 2007). http://dx.doi.org/10.1117/1.2826725

[33] Harriott LR. Next generation lithography. Materials Today. 1999;**2**(2)

[34] Gerald Kreind, Thomas Glinsner, Ron Miller. Next-generation lithography: making a good impression. Nat. Photon. 2010;**4**:27-28. DOI: 10.1038/nphoton.2009.252

Fabrication and Replication of Periodic Nanopyramid Structures by Laser Interference Lithography and UV Nanoimprint Lithography for Solar Cells Applications

Amalraj Peter Amalathas and Maan M. Alkaisi

Additional information is available at the end of the chapter

http://dx.doi.org/10.5772/intechopen.72534

Abstract

In this chapter, the fabrication and replication of periodic nanopyramid structures suitable for antireflection and self-cleaning surfaces are presented. Laser interference lithography (LIL), dry etching, wet etching, and UV nanoimprint lithography (UV-NIL) are employed for the fabrication and replication of periodic nanopyramid structures. Inverted nanopyramid structures were fabricated on Si substrates by LIL and subsequent pattern transfer process using reactive ion etching, followed by potassium hydroxide (KOH) wet etching. The fabricated periodic inverted nanopyramid structures were utilized as a master mold for the nanoimprint process. The upright nanopyramid structures were patterned on the OrmoStamp-coated glass substrate with high fidelity in the first nanoimprint process. In the second nanoimprint process, inverted nanopyramid structures were replicated on the OrmoStamp-coated substrate using the fabricated upright nanopyramid glass substrate as a mold. The replicated inverted nanopyramid structure on resist-coated substrate was faithfully resolved with the high accuracy compared to original Si master mold down to nanometer scale. Both upright and inverted nanopyramid structures can be utilized as surface coatings for light trapping and self-cleaning applications for different types of solar cell and glass surfaces.

Keywords: nanopyramid, laser interference lithography, nanoimprint lithography, reactive ion etching, wet etching, SEM, AFM, thermal evaporation, oxidation, solar cells

1. Introduction

In the last few decades, nanostructure applications have attracted increasing interest in many fields ranging from nanoscale electronics to bionanotechnologies [1–4]. Nanostructures can be

fabricated by several techniques including electron beam lithography (EBL) [5], laser interference lithography (LIL) [6], focused ion beam lithography (FIB) [7], nanosphere lithography (NSL) [8], block copolymer lithography (BCPL) [9], and nanoimprint lithography (NIL) [10, 11].

NIL's ability to provide high-resolution, high-throughput, low-cost, and highly repeatable patterning of nanoscale structures offers potential benefits to numerous electrical, optical, photonic, magnetic, and biological applications. These include hybrid plastic electronics [12], organic laser [13], organic light-emitting diode (OLED) pixels [14], nanoelectronics devices in Si [15], nanoscale protein patterning [16], high-density quantized magnetic disks [17], broad-band polarizers [18], manipulating DNA in nanofluidic channels [19] and solar cells [20–22]. In this study, UV nanoimprint lithography (UV-NIL) process was mainly used to replicate nanopyramid structures for solar cell applications.

This chapter addresses the fabrication and replication of periodic nanopyramid structures by LIL and UV-NIL. Thus, the fabrication processes of the inverted nanopyramid structures on Si substrates by LIL and subsequent pattern transfer process using reactive ion etching (RIE), followed by potassium hydroxide (KOH) wet etching are presented in detail. The development of the UV-NIL and imprint processes for the replication of upright nanopyramid and inverted nanopyramid structures on glass substrates are also discussed in detail.

2. Fabrication of inverted nanopyramid structures

In this section, the fabrication process of periodic inverted nanopyramid structures is discussed step by step. In Section 2.1, the basic theory of LIL and the description of experimental setup details are presented. The design and preparation of the multilayer stack substrates for LIL are described in Section 2.2. In Section 2.3, the details of the single and double LIL exposure process are discussed. Cavity pattern transfer from the soft resist into thermal SiO_2 hard mask layer using reactive ion etching and finally the formation of inverted nanopyramid structures on Si substrate using KOH wet etching are demonstrated in Section 2.4.

2.1. Lloyd's mirror interference lithography

There are various types of interference lithography methods such as Mach-Zehnder interferometer [23], Lloyd's mirror interference lithography (LIL) [24], and scanning beam interference lithography [25], used for different applications. In this work, LIL was utilized to obtain the periodic patterns on photoresist over a large area. The significant benefit of this method is that the period of the pattern can be organized more easily by rotating the stage compared with other two beam interference methods.

A schematic illustration of LIL optical setup is presented in **Figure 1(a)**. A 50-mW HeCd laser with a coherence length of 30 cm at 325 nm was used as the light source in this work. A commercial spatial filter comprises a 5-μm diameter pinhole and a UV objective lens with a focal length of 5.77 mm, which removes the high-frequency noise of the laser to attain a clean Gaussian beam profile. The LIL consists of a substrate holder and an aluminum mirror, both placed perpendicular to each other onto the rotation stage, as shown in **Figure 1(b)**. A UV-enhanced aluminum-coated mirror was used due to its enhanced reflectance in the UV region

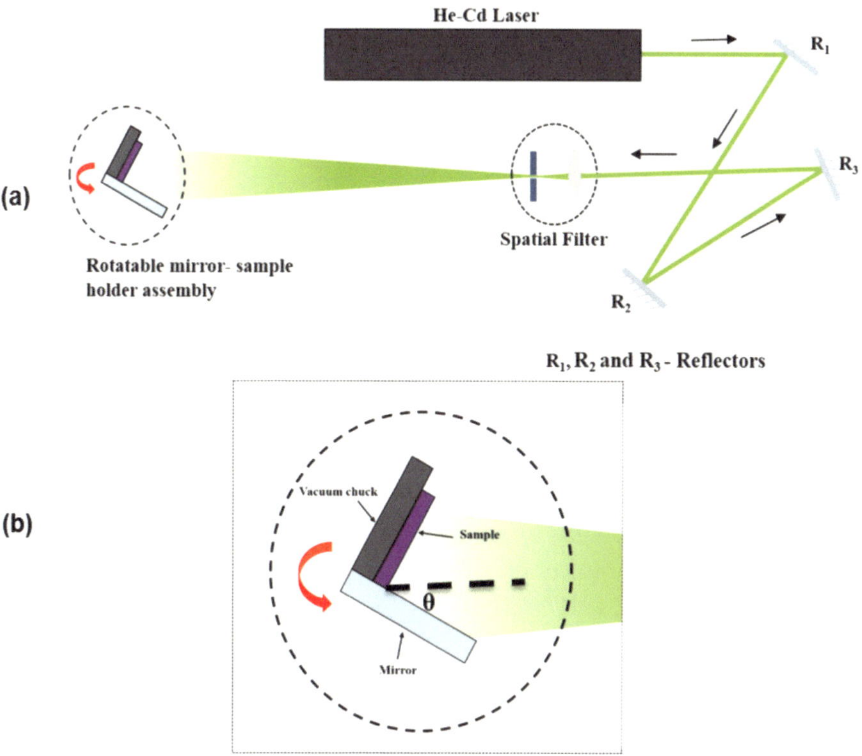

Figure 1. The schematic sketch of optical setup of LIL (a), with the detail of the rotation stage (b).

over a broad range of angles compared to other mirrors. The interference pattern could be disturbed by any vibrations. In order to suppress the vibrations, the complete optical setup was built on the actively damped table. The center of the sample holder and aluminum mirror was located on the optical axis of the laser beam. The coherent laser beam was generated by using the spatial filter at Lloyd's mirror interferometer. A coherent beam illuminates both the mirror and the substrate. There are two parts of the expanded light, which strikes on the substrate. The part of the expanded light, which is directly illuminating the substrate, interferes with the part of the expanded light that is reflected from the mirror surface. This interference gives a periodic line pattern given by Eq. (1). The two-dimensional (2-D) arrays of dots, holes, and variations on them can be recorded with a substrate rotated by 90° during the second exposure. The structure period depends on the laser wavelength and the incident angle between the two beams. The periodicity, p, of horizontal standing wave interference pattern can be represented by Eq. [26]:

$$p = \frac{\lambda}{2\sin\theta} \tag{1}$$

where λ is the laser wavelength and θ is the half angle between the incidence beams. The period of the pattern can easily be controlled by changing the angle θ which is equivalent to the rotation of the stage. The greatest advantage of Lloyd's mirror is the ease of period control. However, the UV-enhanced aluminum-coated mirror's quality (related to flatness and perfections) is a key factor that influences the quality of patterns.

The main reason for the formation of an undesired pattern is the possible presence of the vertical standing wave interference. The primary standing wave formed parallel to the sample, whereas the second standing wave in the vertical direction formed perpendicular to the sample, caused by surface reflection as shown in **Figure 2**. This undesired standing wave is especially present in highly reflecting substrates such as metals or silicon.

The vertical standing wave period is influenced by the factors mentioned in Eq. (1) and the refractive index (n) of the photoresist layer. It is given by Eq. (2) [27]:

$$p_{\text{vertical}} = \frac{\lambda}{2n\cos\theta} \tag{2}$$

The effect of the vertical standing wave can be reduced by decreasing the reflectivity at the interface. In order to prevent these reflections, an extra layer can be added underneath the photoresist layer. This layer should absorb the light and also reflect the light without phase component reflected from the surface. In general, an anti-reflection coating (ARC) is used to suppress the reflections at the interface. Hence, both the refractive index and the thickness of the ARC perform a vital role to cut off the vertical standing wave. To simplify the pattern transfer process, the thin interlayer between the ARC and the photoresist is also used.

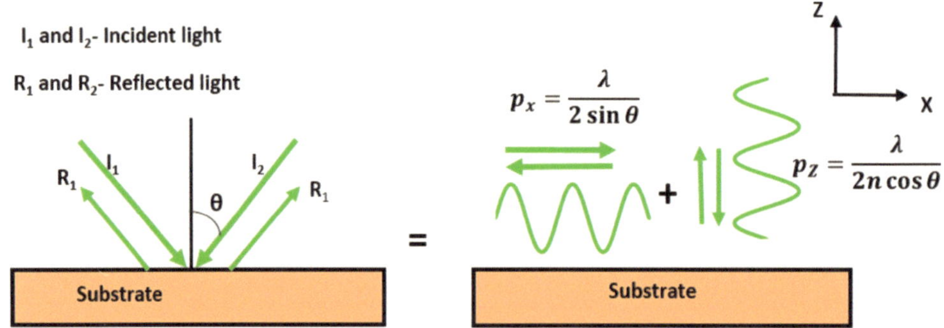

Figure 2. Primary and second standing waves formed parallel and perpendicular to the substrate by respective components.

2.2. Sample preparation

Silicon wafers were primarily used as samples for the LIL processes. A silicon oxide layer was thermally grown on the silicon wafer to act as a masking layer during the wet-etching process, whereas the second silicon oxide layer deposited on the interface between the ARC and photoresist layers using thermal evaporation method for further pattern transfer. The overall process steps involved in the typical sample stack preparation are shown in **Figure 3** for LIL process.

2.2.1. Substrate selection and cleaning

Single-side-polished, Czochralski (CZ) grown, 350-μm thick, Boron-doped p-type silicon wafer with ⟨100⟩ crystal orientation and resistivity of 0.5–1.0 Ωcm was used as substrates. The

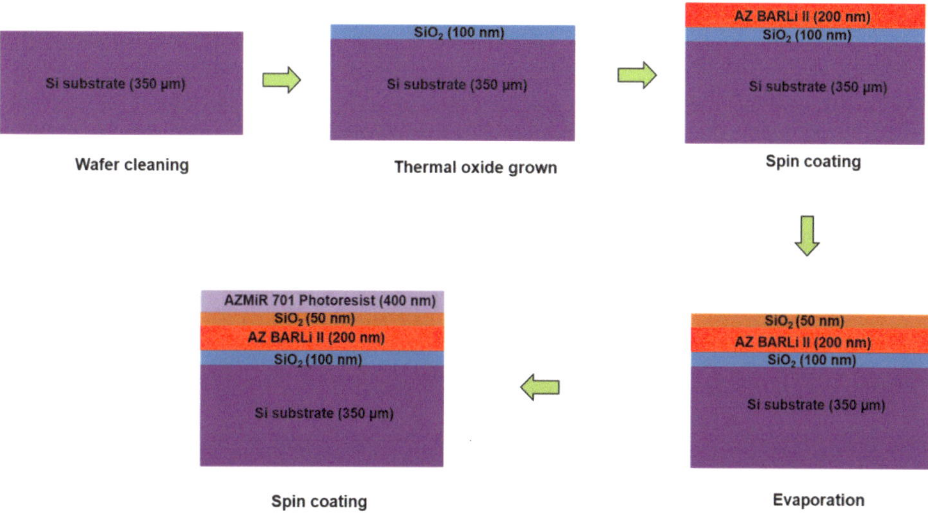

Figure 3. Schematic diagram of multilayer stack preparation process for LIL exposures.

wafer was immersed in a hot mixture of sulfuric acid (H_2SO_4) and hydrogen peroxide (H_2O_2) (3:1 ratio by volume) for 10 min and rinsed with deionized (DI) water [28]. This cleaning procedure was used to remove the metals and organic contamination. Then, the naturally formed silicon oxide layer on the wafer was removed by immersing the wafer in diluted hydrofluoric (HF) acid with DI water (1:10) for 10 s. After that, the wafer was rinsed with DI water and finally blown with nitrogen gas.

2.2.2. *Thermal silicon oxide layer formation*

In order to make the inverted nanopyramid structures on Si, silicon oxide layer was used as a pattern transfer layer and a hard mask during the RIE etching and KOH wet etching without delamination, respectively. A 100-nm thick thermal oxide layer was grown on the cleaned Si wafer using quartz tube furnace. The process parameters are listed in **Table 1**. Oxygen gas was bubbled through the water at 95°C into the oxidation tube to perform the oxidation in a wet oxygen environment. The chemical reaction of the wet oxidation is given by

$$Si(solid) + 2H_2O(vapor) \rightarrow SiO_2(solid) + 2H_2 \tag{3}$$

Parameters	Conditions
Temperature	1000°C
Oxidant species	Wet O_2
Film thickness	100 nm
Oxidation time	12 min

Table 1. Oxidation process parameter.

2.2.3. *Antireflection coating*

The pattern results in the LIL process could be affected by the optical reflections at the interfaces due to the highly reflective silicon substrate. The ARC layer was utilized to reduce the negative effects of the undesirable optical reflections at the interfaces; details of LIL have already been discussed in Section 2.1. For this purpose, AZ BARLi II from MicroChemicals GmbH was used as an ARC resist for the interference lithographic exposures. It is designed to be used with positive photoresist without intermixing. An ARC (AZ BARLi II) 200-nm thick was deposited onto the thermal oxide Si substrate by spin coating at 2250 rpm for 60 s. The sample was then soft baked on a hot plate at 200°C for 60 s to drive the solvent away.

2.2.4. *Evaporated silicon oxide mask layer*

The sample preparation will vary depending on which pattern transfer method will be performed for the fabrication of final hard mold. A bilayer stack consisting of ARC at the bottom layer and photoresist at the top layer is enough for nickel electroplating method, but a trilayer stack with a thin SiO_2 interlayer between ARC and photoresist is essential for pattern transfer. This layer is required to make subsequent reactive ion etching much easier for silicon mold fabrication. It solves selectivity problem between photoresist and ARC during O_2 plasma etch for etching the ARC layer. The 50-nm SiO_2 deposition onto the ARC layer was performed using vacuum thermal evaporation (Balzers BA510A). The base pressure of the chamber was in the range of about 3×10^{-6} mbar and the source material was heated by supplying a high current to the crucible through the molybdenum aluminum oxide boat.

2.2.5. *Photoresist spinning*

In this work, commercially available i-line-positive photoresist (AZMiR 701) was used to record the periodic fringe pattern during LIL process. The undiluted AZMiR 701 resist thickness in the normal spin coating produces thicker layers than the 200–400-nm thickness required. Hence, it was diluted in a ratio of 1:3 with PGMEA (1-methoxy-2-propyl-acetate) to achieve a lower resist thickness. In order to improve the adhesion between the photoresist and SiO_2 layer, an adhesion promoter, hexamethyldisilazane (HMDS, $[(CH_3)_3Si]_2NH$), was employed to deposit a monolayer on the sample surface. After the deposition of the evaporated SiO_2 layer, hexamethyldisilazane was spun onto the substrate, and the 400-nm-diluted positive resist was immediately spin-coated with a 3000-rpm spinning speed for 60 s and soft cooked on a hot plate at 90°C for 1 min to remove any adsorbed moisture. After that, the prepared Si wafer was cut into 2 cm^2 samples.

2.3. Pattern definition using LIL and development

Once the sample stack preparation was complete, Lloyd's mirror setup was performed to pattern the photoresist. The necessary process steps to record the interference patterns were as follows: The prepared Si sample was fixed on the substrate holder rotation stage as shown in **Figure 1**. The rotation stage angle was set as calculated from Eq. (1) for the targeted pattern period. The sample was exposed for the required amount of time using a 50-mW HeCd laser

http://dx.doi.org/10.5772/intechopen.72534

beam operating at 325 nm. A time-controlled shutter was placed between the spatial filter and the rotation stage to control the exposure time during each exposure. The line pattern on photoresist was recorded by the single exposure. Holes, dots, and variations of patterns were recorded by a double exposure with a sample rotated by 90° after the first exposure.

After the exposure, the exposed sample was immersed and carefully agitated in diluted Microchemicals AZ MIF 326 developer solution for 30 s. At this stage, the exposed part of the photoresist was dissolved, leaving the required pattern on the photoresist. The sample was rinsed immediately with deionized water and finally blown with nitrogen gas. The developed samples were examined by scanning electron microscope (SEM). More details will be presented in Sections 2.3.1 and 2.3.2.

2.3.1. Single exposure pattern

For positive photoresist use in LIL, in a single exposure and development step, periodic line grating pattern will be produced. In this section, a significant parameter of the LIL process the

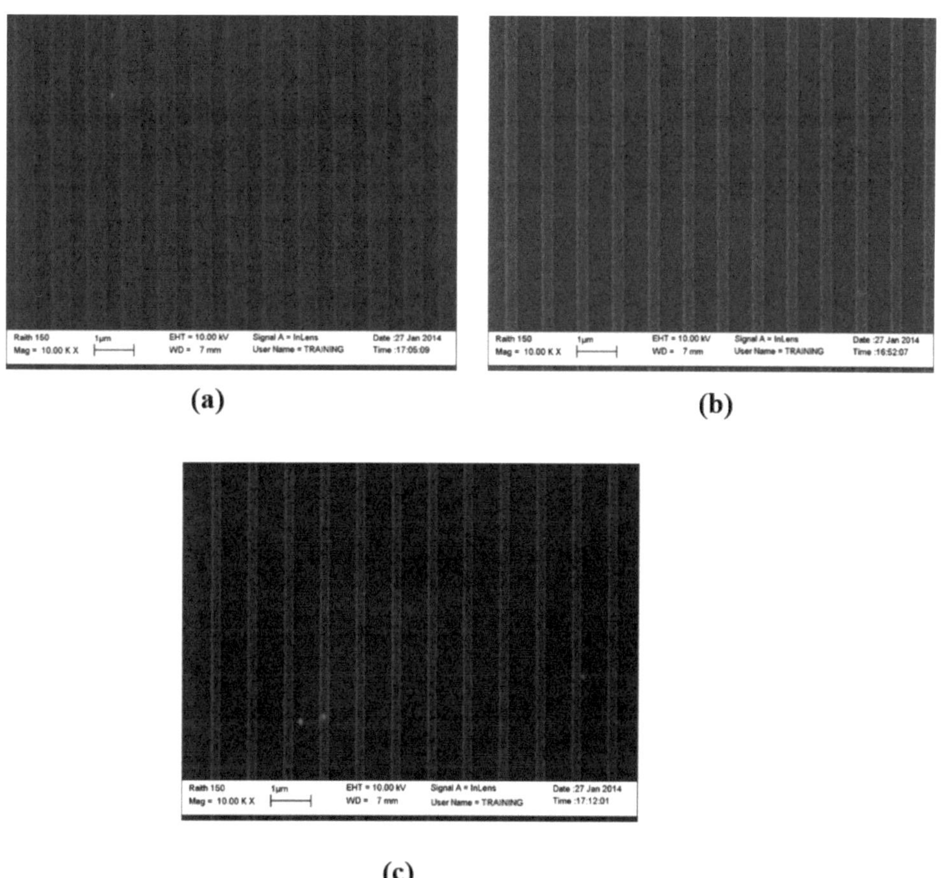

Figure 4. SEM images of 900-nm period line grating pattern on photoresist with different line widths and exposure time: (a) 520 nm and 120 s, (b) 310 nm and 240 s, and (c) 250 nm and 360 s.

so-called "duty-cycle" (DC), is presented in order to evaluate the exposure results. The DC is defined as the ratio of the pattern line width (W_{line}) generated by LIL to the period ($P_{grating}$) of the pattern and is represented by Eq. (4). As described in Eq. (1), the period of the pattern relies on the laser wavelength and the laser beam incident angle

$$DutyCycle(DC) = \frac{W_{line}}{P_{grating}} \tag{4}$$

In this case, the laser wavelength (325 nm) is a constant value. The pattern line width value could be modified by changing the DC value for a fixed period, which relies on the exposure dose. The exposure dose (D) in energy per unit area was calculated by multiplying the time of exposure (t) and the incident laser beam intensity (I_0). It can be concluded that the exposure time and the intensity of the laser beam can impact the exposure results. The intensity of the incident laser light at a fixed incident angle could be considered as a constant value. Thus, the exposure dose is directly proportional to the exposure time. The DC value can be controlled by varying the exposure time at a fixed incident angle.

Figure 4 shows the SEM images of the 900-nm period line grating pattern with a different exposure time of 120, 240, and 360 s. In this case, the intensity of exposure (0.15 mW) and the rotating angle of the sample holder ($\theta = 10^{\circ}24'$) remained constant. As shown in **Figure 4**, the pattern line width/DC value decreased by increasing the exposure times at a fixed incident angle and intensity. Therefore, it can be concluded that the exposure time can impact the final pattern results at a fixed incident angle.

For LIL exposures for different periods, one cannot merely assume a constant exposure time, even under identical exposure conditions. As shown in **Figure 5**, the light density on the sample surface at a normal incidence is higher than that of oblique incidence due to increasing

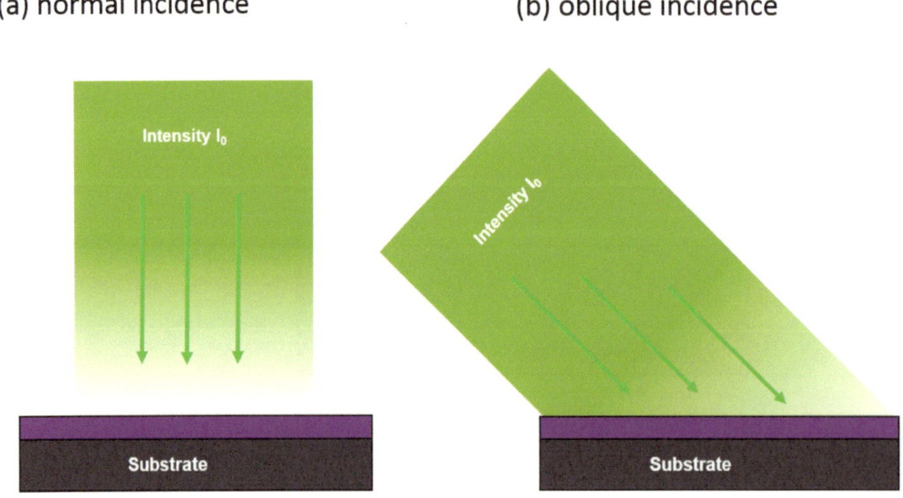

Figure 5. Schematic illustration of the exposed area at (a) normal incidence and (b) oblique incidence for fixed laser intensity. The sample area illuminated by the beam is smaller at normal incidence.

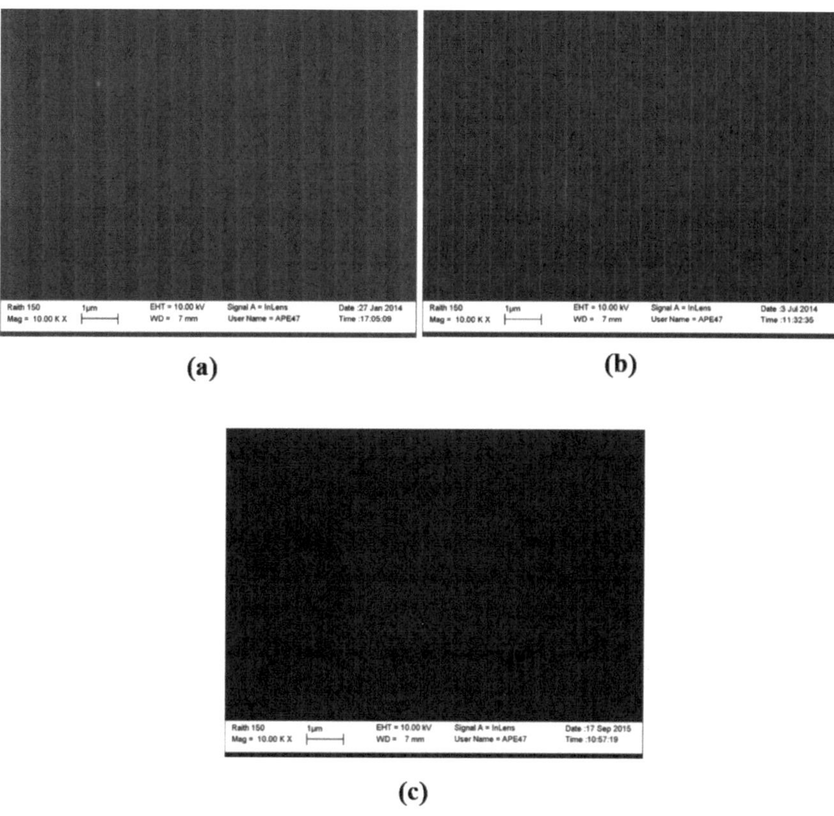

Figure 6. SEM images of line grating pattern on photoresist with different periods and exposure time: (a) 900 nm and 120 s, (b) 700 nm and 150 s, and (c) 300 nm and 270 s.

the exposed area at oblique incidence angles. As described in Eq. (1), the incident angle for a larger period is smaller than that for a smaller period. It can be deduced that longer exposure times are required for smaller structure period.

In this work, well-ordered experiments were employed in order to find the relationship between the angle of incidence and exposure time. **Figure 6** shows the resulting SEM images of line grating pattern having 900-, 700-, and 300-nm periods with the exposure time of 120, 150, and 270 s, respectively. In order to achieve periods of 900, 700, and 300 nm, the angles of incidence were adjusted to $10°24'$, $13°25'$, and $32°47'$, respectively. The exposure time has to increase from 120 to 270 s by using a higher angle of incidence. Thus, larger incident angle (smaller period) requires longer exposure times.

2.3.2. *Double exposure pattern*

For positive photoresist LIL system, after the double exposure and development, periodic holes or dots will be obtained. **Figure 7** shows the developed pattern of resist holes and dots with a period of 700-nm square array. The patterns were formed by double exposure IL with the sample rotated by 90° between the two exposures. Exposure times of 70–120 s with a 10-s increment for each experiment were used. In order to achieve the period of 700 nm, the angle of incidence was adjusted to 13°25′. As explained in the previous section, exposure dose depends mainly on the exposure time for constant laser intensity at a fixed incident angle.

Even at the lowest exposure dose, a pattern of holes was formed as shown in **Figure 7(a)**, although there was a little variation in the size of holes. With increasing exposure dose, the size of the holes increased and became more uniform as shown in **Figure 7(b)** and **(c)**, then holes pattern vanished but some dots pattern still linked together as shown in **Figure 7(d)**. The exposure dose further increases results in an isolated dot array as shown in **Figure 7(f)**. It can be seen that exposure time can significantly impact the size and type of pattern structures. In

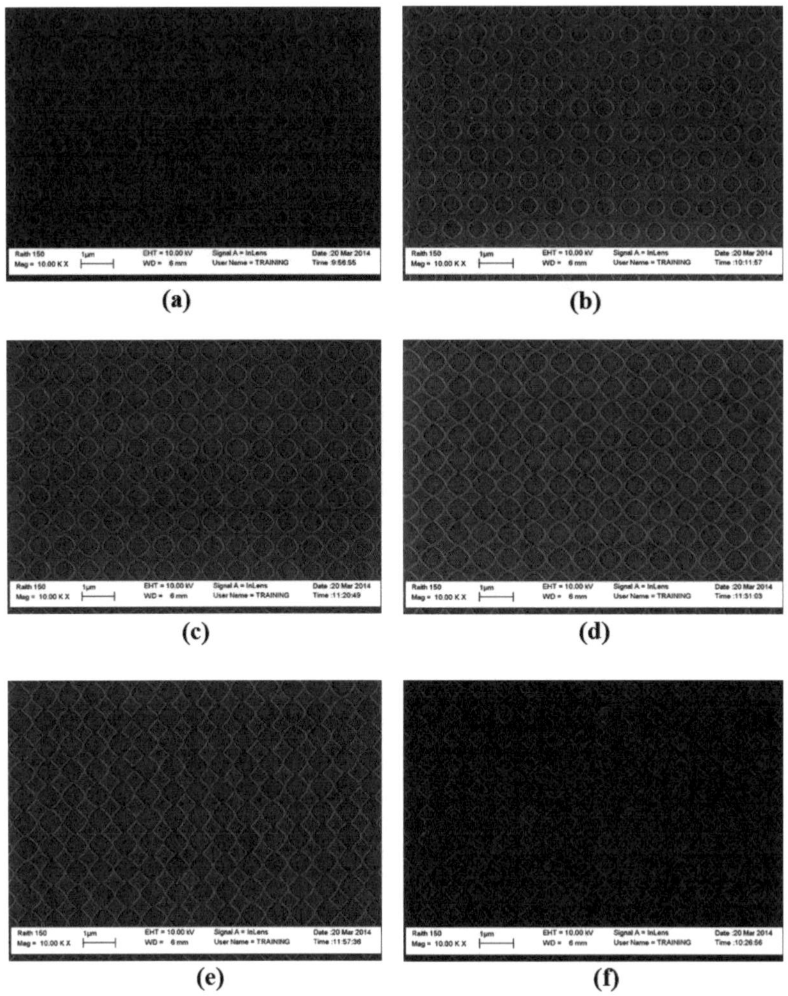

Figure 7. SEM images of holes and dots pattern on photoresist produced by double exposure IL with a period of 700 nm and exposure times of (a) 70, (b) 80, (c) 90, (d) 100, (e) 110, and (f) 120 s for each exposure. The samples were rotated by 90° between the two exposures.

order to fabricate the inverted nanopyramid on Si substrate, a pattern of resist holes is required for this task. More details about the pattern transfer and formation of inverted nanopyramid will be discussed in the next section.

2.4. Pattern transfer and formation of inverted pyramid

The periodic inverted nanopyramid structures were fabricated on Si substrate by laser interference lithography and subsequent pattern transfer process using reactive ion etching, followed by KOH wet etching. The schematic illustration of the fabrication process of inverted nanopyramid structure on Si substrate is shown in **Figure 8**. After the LIL patterning of photoresist, the resist pattern of holes was transferred onto the bottom SiO_2 layer by a subsequent reactive ion etching process. Then, the inverted pyramid structures were formed by KOH wet etching. KOH has anisotropic etching profile with a selectivity of 400:1 to ⟨100⟩:⟨111⟩ orientations in silicon. Finally, the thermal oxide layer was removed by buffered hydrofluoric etching. More details of the fabrication process will be discussed in the next section.

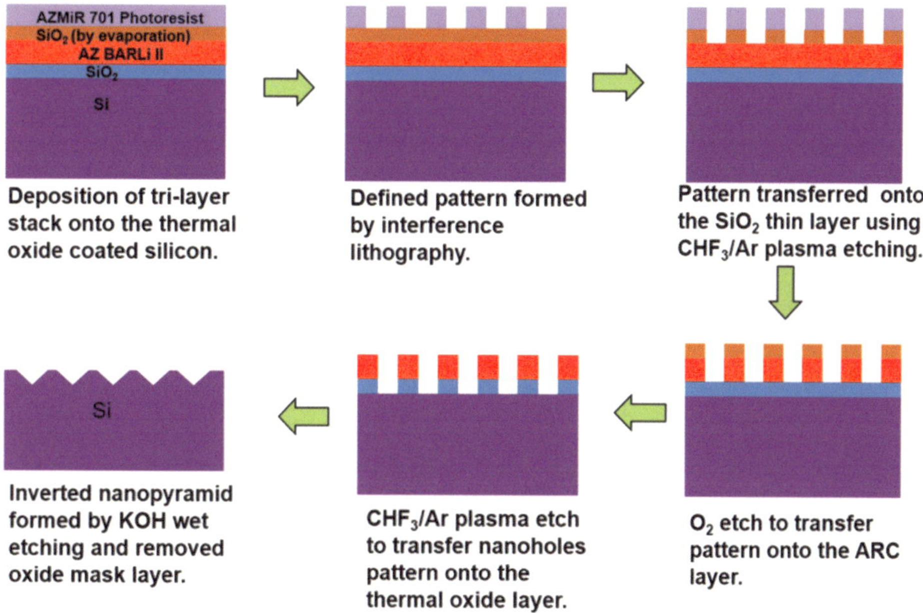

Figure 8. Schematic illustration of the fabrication process of inverted nanopyramid structures on an Si substrate.

2.4.1. *Dry plasma etching*

In this work, an Oxford PlasmaLab80 reactive ion etching system was utilized for all the dry plasma etching steps. The RIE etching process parameters such as flow rate of the processing gases, pressure, substrate temperature, RF power, and etching time were carefully optimized for the pattern transfer to form the master mold.

2.4.1.1. *Pattern transfer into silicon oxide layer*

The resist pattern produced by LIL served as the etching mask for the RIE pattern transfer step into the thermally evaporated silicon oxide layer. Generally, an ARC layer is etched slower than the photoresist layer. Hence, the resist pattern of holes produced by LIL cannot be directly transferred into the thermal oxide layer. The thin evaporated SiO_2 layer was deposited between the ARC and the photoresist in order to get a high-etching selectivity. Before the SiO_2-etching process, the residual photoresist at the bottom of the holes was removed using a little O_2 plasma etching. The CHF_3/Ar plasma etching was performed to transfer the pattern of holes onto the thin SiO_2 interlayer. The RIE process parameters for O_2 plasma for descumming the residual resist and transferring the pattern into an SiO_2 layer using CHF_3/Ar plasma is illustrated in **Table 2.**

The SEM image of the developed resist pattern of holes with a period of 650-nm square arrays is shown in **Figure 9(a)**. In order to achieve a period of 650 nm, the angle of incidence was adjusted to 14°28′. **Figure 9(b)** displays an SEM image of the pattern of holes transferred into an SiO_2-masking layer. It can be seen that the pattern uniformity remains very high.

RIE parameters	Removal of residual resist layer	Pattern transfer into SiO_2 layer
Gas	O_2	CHF_3/Ar
Flow rate	10 sccm	25/30 sccm
RF power	100 W	150 W
Pressure	100 mTorr	100 mTorr
Temperature	295 K	300 K
Time	10 s	1 min 30 s
Masking material	—	Photoresist

Table 2. RIE recipes for the removal of residual resist layer and transferring the holes pattern into the SiO_2 layer.

2.4.1.2. *Pattern transfer into ARC layer*

The SiO_2 interlayer acted as a masking layer to transfer the holes pattern into the ARC layer. The O_2 plasma etching is performed to transfer the pattern into ARC layer. **Table 3** outlines the optimized RIE parameters of O_2 plasma etching for transferring the patterns into ARC layer. **Figure 10** shows the SEM image of patterned ARC layer after the O_2 plasma etching.

2.4.1.3. *Pattern transfer into thermal oxide layer*

The patterned ARC layer served as a masking layer to transfer the holes structure into a thermal SiO_2 bottom layer. Then, CHF_3/Ar plasma etching should be performed to transfer the pattern into ARC layer. **Table 4** shows the optimized RIE parameters for CHF_3/Ar plasma etching for transferring the patterns into an SiO_2 bottom layer. **Figure 11** shows the SEM image of patterned SiO_2 bottom layer after the CHF_3/Ar plasma etching. It can be observed that the RIE-etching process induced a slight enlargement of holes diameter while the uniformity was improved.

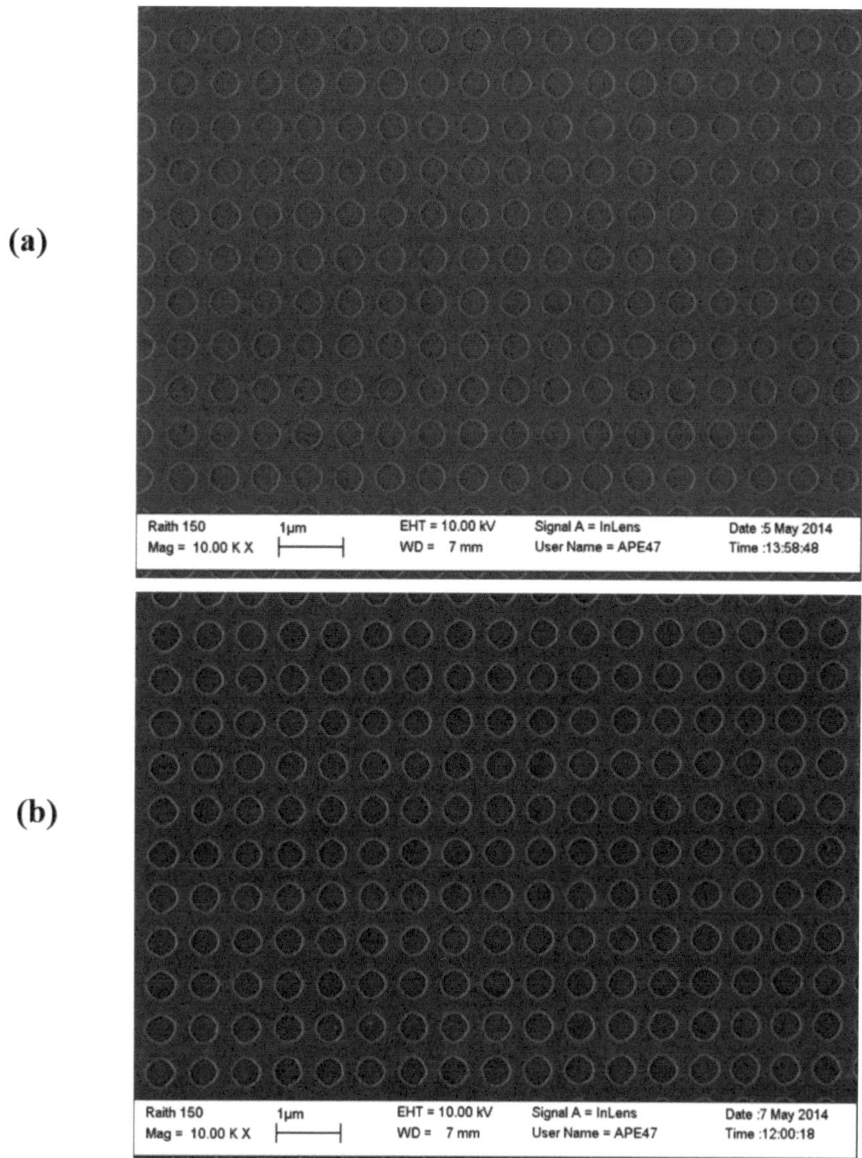

Figure 9. SEM images showing (a) the developed patterns on resist with a period of 650 nm and (b) patterns transferred into SiO_2-masking layer after CHF_3/Ar plasma etching.

2.4.2. *Wet chemical anisotropic etching*

After the RIE process, the next process step was anisotropic wet chemical etching in order to form inverted pyramid structures into the silicon substrate. The anisotropic KOH chemical etching characteristic of single crystal silicon substrate varies according to the crystallographic orientation of the substrate bulk material. Anisotropic etchants such as potassium hydroxide, sodium hydroxide (NaOH), cesium hydroxide (CsOH), ammonium hydroxide (NH_4OH),

RIE parameters	Transferring the pattern into ARC layer
Gas	O_2
Flow rate	10 sccm
RF power	100 W
Pressure	100 mTorr
Temperature	300 K
Time	13 min 30 s
Masking material	SiO_2 layer

Table 3. The optimized O_2 plasma RIE parameters for pattern transfer into ARC layer.

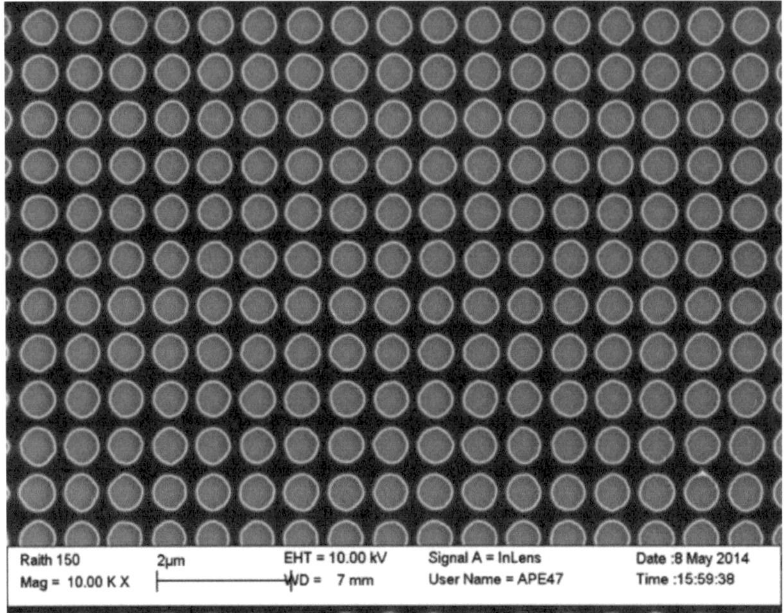

Figure 10. SEM images of patterns transferred into ARC layer after O_2 plasma etching.

RIE parameters	Transferring the pattern into SiO_2 bottom layer
Gas	CHF_3/Ar
Flow rate	25/30 sccm
RF power	150 W
Pressure	100 mTorr
Temperature	300 K
Time	3 min 30 s
Masking material	ARC layer

Table 4. The optimized CHF_3/Ar plasma RIE parameters for pattern transfer into the thermal SiO_2 bottom layer.

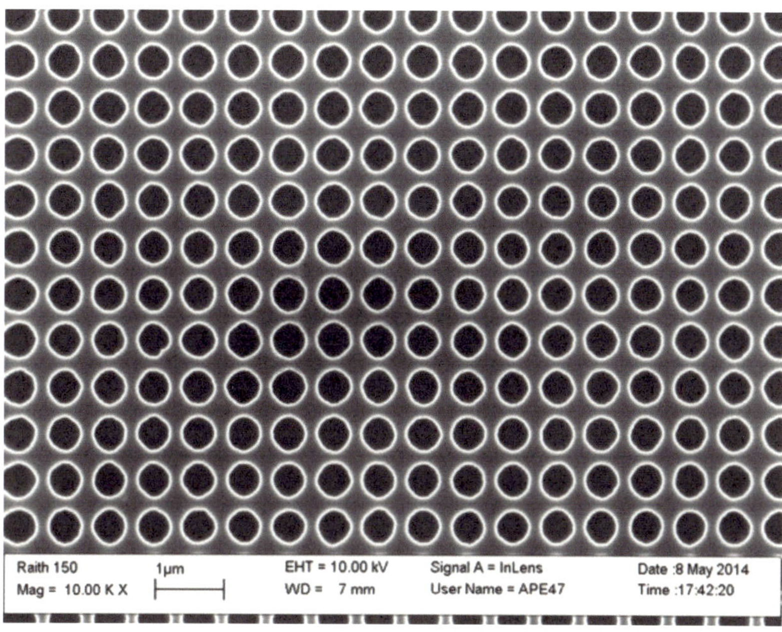

Figure 11. SEM images of patterns transferred into SiO_2 layer after *CHF_3/Ar* plasma etching.

tetramethylammonium hydroxide (TMAH), and hydrazine etch specific crystallographic planes at different etch rates. The etch rate of the (111) plane is significantly low compared to the (100) plane etch rate mainly due to some dangling bonds in each unit cell [29]. The (111) plane is denser and has the lowest dangling bond; more back bonds must be broken and therefore the etch rate is low. The etch rates for anisotropic wet chemical etchants are mainly

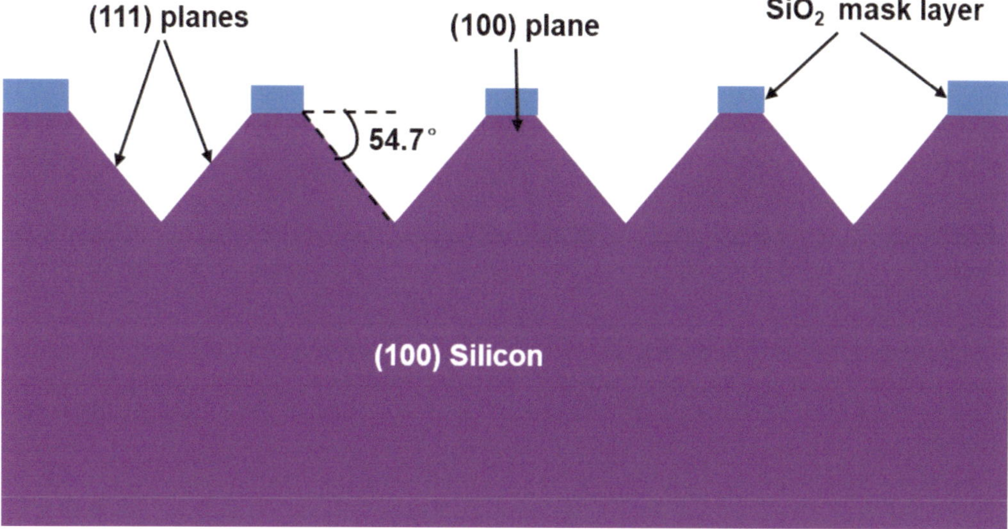

Figure 12. An anisotropic wet etch on a (100) silicon substrate creates inverted nanopyramid structure.

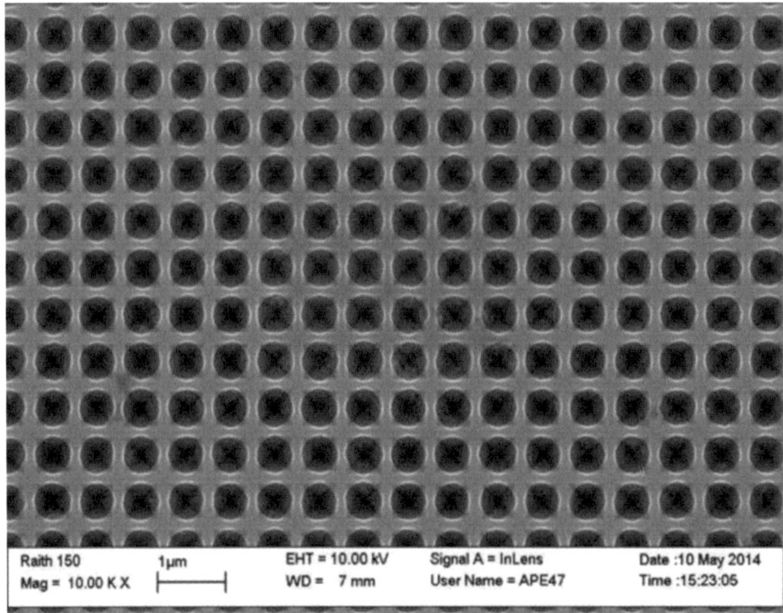

Figure 13. SEM images of inverted nanopyramid structures on Si substrate with SiO_2-masking layer after KOH wet etching.

dependent on the concentration of the solution and temperature. As schematically illustrated in **Figure 12**, the inverted nanopyramid structure could be produced with ⟨100⟩ crystalline orientation of silicon substrate using anisotropic wet etching. Anisotropic etchants make an angle of 54.7° for the ⟨100⟩-oriented Si substrate, which is the angle between (100) and (111) planes.

The chemical etchant for anisotropic wet etching used in this work is typically a 30 wt.% KOH diluted in deionized water. SiO_2 was chosen as a hard mask layer mainly due to the high-etch selectivity of KOH solution between Si and SiO_2 about 100:1. The inverted pyramid structure is formed on an (100) Si substrate by wet etching in 30% KOH in deionized water solution at 80°C for 170 s. The KOH solution was maintained in a temperature-controlled bath at 80°C. **Figure 13** shows the SEM image of the resulting inverted nanopyramid structure having a width of about 600 nm and separation of about 100 nm, obtained from the same sample as in **Figure 9**.

2.4.3. *Removal of silicon oxide mask layer*

There were at least two techniques available for the removable of the SiO_2 mask layer, wet etching using buffered HF solution or dry etching using CHF_3/Ar plasma. The wet etching process was used in this work to avoid increased surface roughness or any surface damage. To remove the SiO_2 mask layer from the patterned substrate, the substrate was immersed in buffered HF (6:1 volume ratio of NH_4F solution to 49% HF) for 3 min, washed with DI water, and blown dry with N_2. The SEM image of the formed periodic inverted nanopyramid

http://dx.doi.org/10.5772/intechopen.72534

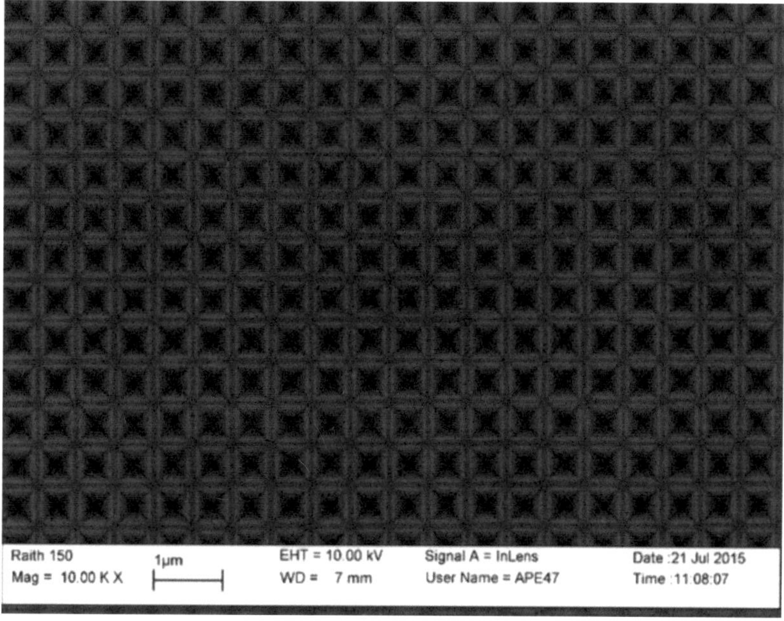

Figure 14. SEM images of inverted nanopyramid structures on Si substrate after removal of the SiO_2-masking layer.

structures on Si substrate after removal of the SiO_2-masking layer is shown in **Figure 14**. It can be seen that all the inverted pyramid structures were completely formed and centered without showing any overlapping between neighbors. It can also be observed that the size of the inverted pyramid increased slightly compared with a diameter of the nanoholes as in **Figure 13** due to undercutting during KOH wet etching. The fabricated inverted nanopyramid structures on Si Substrate were utilized as a master mold in the nanoimprint replication process. The replication process of the inverted nanopyramid structures will be introduced in Section 3.

3. UV nanoimprint lithography: replication of nanopyramid structures

Nanoimprint lithography was first proposed and demonstrated by Chou et al. in 1995 alternative to e-beam lithography and photolithography as a low-cost and high-throughput technique [30]. NIL can generally be classified into two fundamental categories: hot-embossing lithography (HEL) also known as thermal nanoimprint lithography (TNIL) and UV-NIL. Currently, many different kinds of lithographic methods were established based on NIL, typical examples include roll-to-roll imprint lithography [31], step-and-flash imprint lithography (SFIL) [32], laser-assisted NIL [33], microcontact printing [34], reverse imprint lithography [35], and step-and-stamp imprint lithography (SSIL) [36].

NIL is based on the mechanical deformation of the resist using a mold consisting of micro- or nanostructures by UV-curing process or thermomechanical process. The UV-NIL method has several benefits compared to the thermal NIL, which includes the capability of UV-NIL to be performed at room temperature. The imprinting process at room temperature eliminates the problems resulting from thermal expansion mismatch between the sample, resist, and mold [37]. Moreover, the UV-NIL-imprinting process contains a lower viscosity of the photoresist, which permits the imprinting process to be performed at a lower pressure in comparison with thermal NIL [38, 39]. The cycle time of the UV-NIL process is shorter than the thermal NIL due to the elimination of temperature cycle, which improves the process throughput [40]. However, the UV-NIL process requires either a transparent substrate or a transparent mold. In this chapter, UV-NIL process was mainly used to replicate nanopyramid structures for solar cell applications.

In this section, the development of the ultraviolet nanoimprint lithography and imprint processes for the replication of upright nanopyramid and inverted nanopyramid structures are presented. First, the sample preparation processes for UV-NIL are introduced in Section 3.1. Then, imprint processes for the replication of upright nanopyramid and inverted nanopyramid structures and patterns analysis are described in Section 3.2.

3.1. Preparation for imprint process

The materials required for UV-NIL are a mold consisting of the fabricated nanostructures and an appropriate UV-curable resist that could be deformed and hardened to reproduce the shape of the structures. In this section, the anti-sticking layer preparation of the master mold and UV-curable resist material preparation on a glass substrate for imprint process are presented.

3.1.1. Master mold fabrication

The mold is one of the essential components of the UV-NIL process as it contains the pattern information and details. The solid materials with a high hardness and durability properties can be used as a mold to NIL. A variety of materials including silicon, silicon nitride, silicon dioxide, quartz, glass, nickel, and so on have been used to make molds to NIL. In this chapter, the periodic inverted nanopyramid structure on the silicon substrate was used as a master mold substrate for the imprint process. The periodic inverted nanopyramid structures were fabricated on Si substrate by LIL and subsequent pattern transfer process using reactive ion etching, followed by KOH wet etching. The details of the fabrication process of inverted nanopyramid structures are described in Section 2.

3.1.2. Antisticking layer treatment of mold

A master mold surface to NIL has a high density of nanoscale structures, which efficiently enhances the surface area. The mold with increased surface area contacts with the imprinted resist, leading to high adhesion of the mold to the resist. This adhesive effect could be observed by the sticking of the imprinted resist to the mold without any anti-sticking surface treatment. The anti-sticking treatment for NIL molds eliminates the adhesion between the mold and imprint resists which enhances the qualities of the imprint. Moreover, it also improves the mold lifetime remarkably by avoiding surface damage and contamination. Hence, it is necessary to deposit an anti-sticking layer directly onto the master mold before the imprint process. The SAM coating with low-surface energies such as silane materials [41] and Teflon [42] has been reported as an anti-sticking layer to enhance the demolding abilities and to enhance the

http://dx.doi.org/10.5772/intechopen.72534

mold lifetime. The most widely used anti-sticking layer approach is a SAM of a fluorosilane release agent by either a vapor phase or a solution phase deposition.

In this work, a 1H, 1H, 2H, 2H–perfluorooctyl-trichlorosilane (F_{13}-TCS) solution was utilized as an anti-adhesive coating on the mold for NIL. An anti-sticking coating was formed on the mold surface with F_{13}-TCS agent via a vapor deposition method inside the desiccator at room temperature. The mold was cooked in the oven at 90°C for 30 min to completely dehydrate the mold surface and then cooled down to room temperature. A few drops of the F_{13}-OTCS solution were added to a small Petri dish which was loaded with mold into a vacuum desiccator as shown in **Figure 15**. The substrate was left to react for 2 h at room temperature, then removed from the desiccator and baked in the oven at 90°C for 1 h.

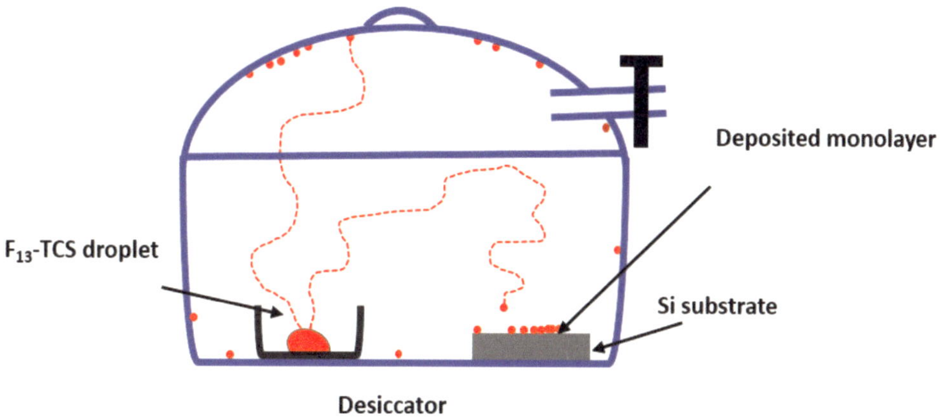

Figure 15. Schematic representation of the anti-adhesive coating using a vapor deposition method of F_{13}-TCS in a vacuum desiccator.

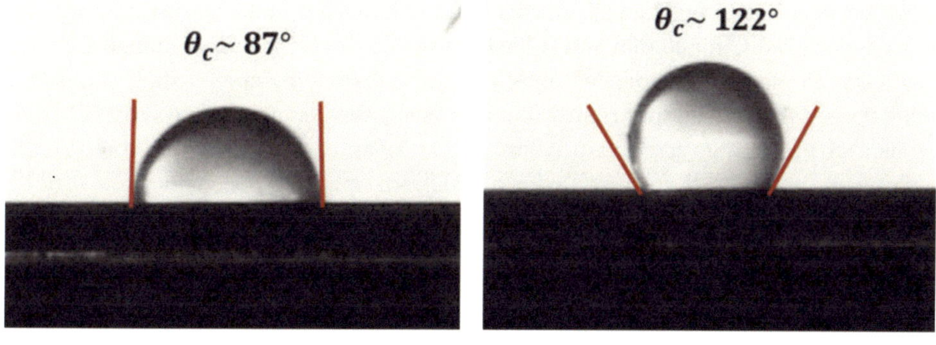

(a) Inverted nanopyramid structured Si substrate.

(b) Inverted nanopyramid structured Si substrate with SAM.

Figure 16. The contact angles on the surface of inverted nanopyramid-structured Si substrate before and after the anti-sticking layer treatment.

Prior to the imprint process, the contact angle measurements were performed on the inverted nanopyramid-textured Si substrates before and after the anti-sticking layer treatment to quantify the variations of the surface energies. The contact angle of inverted nanopyramid-textured Si surfaces before and after an anti-sticking layer treatment is shown in **Figure 16**. The contact angle of the inverted nanopyramid-textured Si surface was increased from 87 to 122° after coated with an anti-sticking layer of F_{13}-TCS which is highly hydrophobic. It can be clearly seen that F_{13}-TCS SAM coating was enhanced by the anti-adhesive properties on the inverted nanopyramid-textured Si substrate which is crucial for the imprinting process.

3.1.3. Substrate preparation

The low iron glass with a thickness of 0.5 mm was used as a substrate for UV-curable resist. The UV-curable resist plays a vital part in the efficient nanoimprint process where sticking properties between the mold and UV curable polymer should be as low as possible, whereas the sample and resist should be high [40]. The resist that was used in this work was OrmoStamp from Micro Resist Technology, which is UV-curable resist with high transparency for UV and visible light designed for UV imprinting or molding.

The OrmoStamp resist must have a strong interfacial bonding to the substrate, but not stick to the surface of the mold in the imprint process [43]. In order to obtain a strong adhesion between the glass substrate and resist, the cleaning process with oxygen plasma treatment was performed, and also adhesion promoter was added on the glass substrate prior to the resist spin coating. The OrmoPrime08 from micro resist technology was used as an adhesion promoter solution based on organofunctional silanes. It has been designed to promote the adhesion of OrmoStamp, OrmoComp, OrmoClear, OrmoCore, and OrmoClad to various substrates like silicon, glass, and quartz.

The glass substrate was cleaned with acetone, methanol, and isopropyl alcohol (IPA) solvents in an ultrasonic bath and then rinsed with deionized water and finally dried with nitrogen gas. Next, the substrate surface was treated with a short time O_2 plasma to enhance optimum adhesion between the OrmoPrime08 and glass interface. After that, the substrate was baked using an oven at 200°C for 30 min and cooled down to room temperature immediately before coating. OrmoPrime08 was deposited onto the glass substrate by spin coating at a 4000 rpm spinning speed for 60 s. The spin-coated film was then baked on a hot plate at 150°C for 5 min and cooled down to room temperature. Finally, OrmoStamp resist was spin-coated onto the OrmoPrime08 layer-coated substrate with a 6000-rpm spinning speed for 60 s. After spin coating, the substrate coated with resist was thermally prebaked on a hot plate at 80°C for 2 min to enhance the uniformity of the resist thickness and to increase the adhesion between the resist and the substrate.

3.2. Nanoimprint process

In this work, two vacuum-operated in-house built imprint tools were utilized to perform the imprint experiments on 20×20 mm^2 and 10×10 mm^2 samples, respectively. It creates a vacuum region between the resist and the mold to decrease the air bubbles surrounded in between them during the imprint process.

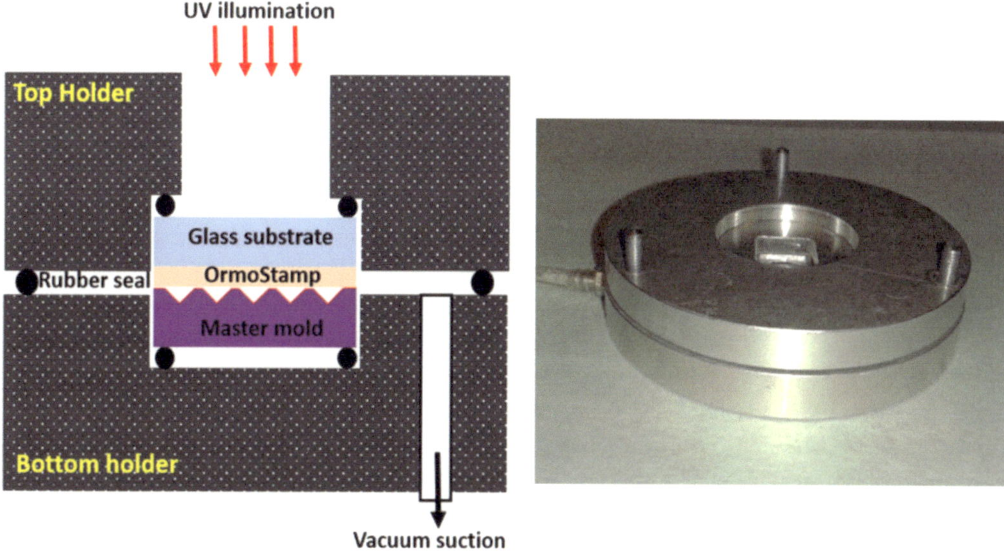

Figure 17. The vacuum-operated in-house built imprint tool utilized for UV-NIL.

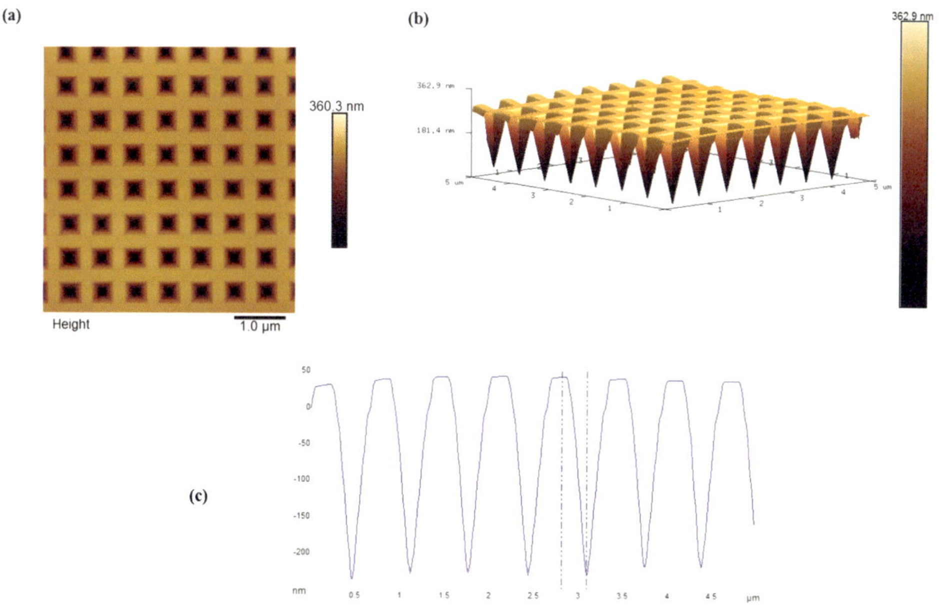

Figure 18. AFM images of inverted nanopyramid structured master mold: (a) 2-D view, (b) 3-D view, and (c) cross-sectional traces.

Figure 17(a) and **(b)** show the schematic cross-sectional view and the optical image of the imprint tool used in this work, respectively. The Mask Aligner (MA-6) exposure system was attached to this imprint tool to perform as a UV-NIL tool. The Mask Aligner system uses i-line 365-nm wavelength UV source for exposure. The Mask Aligner vacuum system is linked to the imprint tool that is used to hold the mask onto the imprint holder. The 365-nm wavelength UV illuminations with a vacuum pressure of 4 mbar were employed to perform the imprint process.

In order to replicate the original master mold, two subsequent imprint processes need to be applied. In this work, periodic inverted nanopyramid-textured Si substrate was used as a master mold. In the first step of the imprint process, the negative of the pattern on the master mold was replicated from Si master mold onto resist-coated glass substrate, that is, inverted nanopyramid structures on Si master mold become upright nanopyramid structures on resist-coated glass substrate. In the second step of imprint process, the inverted nanopyramid structures were replicated on resist using upright nanopyramid-patterned glass substrate as a master stamp. **Figure 18** shows the atomic force microscopy (AFM) images and cross-sectional traces of periodic inverted nanopyramid structures on Si. The periodic inverted nanopyramid structures have features with a width of about 450 nm, a height of about 300 nm, and separation of about 150 nm.

3.2.1. *Imprint: Upright nanopyramid structures replication*

Figure 19 shows the schematic illustration of the first imprint process steps to create the upright nanopyramid structures into a UV-curable resist-coated glass substrate from inverted nanopyramid-structured Si master mold. The F_{13}-TCS SAM-coated Si master mold coated with F_{13}-TCS SAM/UV-curable resist-coated glass substrate was loaded into the imprint tool. The SAM-coated Si master mold and the UV-curable-coated glass substrate were prepared as described in Section 3.1. A vacuum pressure was set to 4 mbar and the Mask Aligner (MA-6) system was then activated. The resist was cured under a UV exposure for 4 min using 4.4 mW/cm^2 illumination intensity with 365-nm UV source at room temperature. A manual detaching process was utilized by giving gentle force using a blade at one edge of the substrate in order to delaminate between the mold and the substrate surfaces. Subsequently, the replicated substrate was thermally baked in an oven at 150°C for 2 h to improve the film thermal and environmental stability.

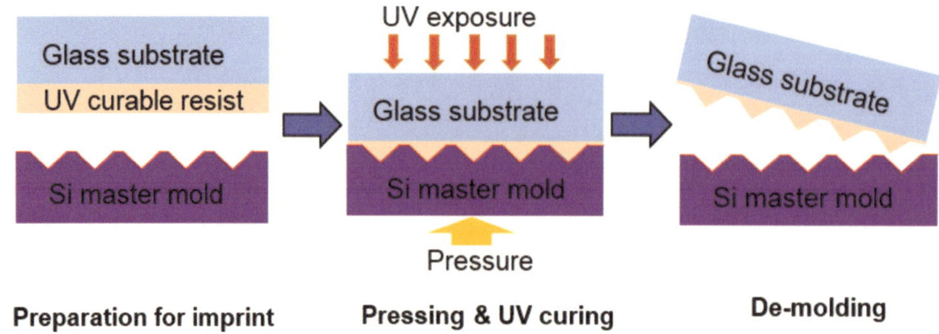

Figure 19. The schematic diagram of the first imprint process steps to replicate the upright nanopyramid structures.

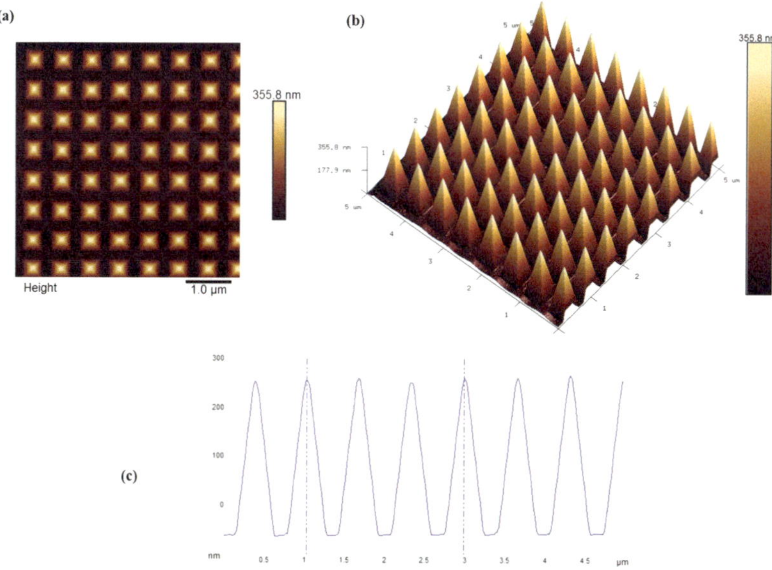

Figure 20. AFM images of the upright nanopyramid structures formed on UV transparent OrmoStamp resist-coated glass substrate after the first imprint: (a) 2-D view, (b) 3-D view, and (c) cross-sectional traces.

Figure 20(a) and (**b**) show the 2-D and three-dimensional (3-D) AFM images of the periodic upright nanopyramid formed onto the OrmoStamp resist-coated glass substrate as the result of first imprint process. It can be seen that the upright nanopyramid structures with periodic features in the order of 500 nm and smooth surfaces have been precisely replicated onto the OrmoStamp resist-coated glass substrate. The AFM images for the master mold and imprinted sample were compared and dimensions measured from randomly selected areas but with the same scanned area of 5×5 μm. **Figures 18(c)** and **20(c)** reveal that no significant differences can be found between the master mold and inverted shape of master mold replica. These results further confirm that excellent fidelity periodic upright nanopyramid structures can be achieved by UV-NIL imprinting. This high-fidelity replication offers high flexibility in designing new light-trapping schemes for solar cell applications. The UV-curable resist can be incorporated into a range of solar cell configuration because of its low optical absorption [44]. Therefore, the replicated periodic upright nanopyramid structures on the OrmoStamp resist-coated glass substrate can be utilized as light-trapping and self-cleaning functions in different types of solar cells such as thin film and polycrystalline materials. This imprinting process can be continued, and the inverted nanopyramid structures can be created from the upright nanopyramid-structured glass as a mold for a direct 3-D imprint process. It should be noted that there is no direct technique for forming periodic and ordered upright pyramid structures on crystalline silicon because of the limitation imposed by the crystal orientation.

3.2.2. Imprint: inverted nanopyramid structures replication

Figure 21 shows the overall imprinting process steps to create the inverted nanopyramid structures onto UV-curable resist-coated glass substrate. In this imprinting process, the replicated upright nanopyramid structure on UV-curable resist-coated glass substrate was used as a mold in order to form the inverted nanopyramid structures on UV-curable resist-coated glass substrate. A very thin F_{13}-TCS SAM was used as an anti-adhesive layer on the upright nanopyramid-structured glass substrate. It was deposited on a resist-coated glass surface as described in Section 3.1.2.

The OrmoStamp-coated substrate was prepared as explained in Section 3.1.3 and the upright nanopyramid-structured glass substrate coated with F_{13}-TCS SAM/the OrmoStamp-coated substrate was loaded into the imprint tool as shown in **Figure 21**. The inverted nanopyramid structures were replicated on the OrmoStamp-coated substrate by the same UV-imprinting method as described in the first imprinting process using upright nanopyramid-patterned glass substrate as a master mold.

In order to determine the accuracy of the replication by the nanoimprint process, AFM images of the master mold and replicated substrates were taken. **Figure 22(a)** and **(b)** show the 2-D and 3-D AFM images of the periodic inverted nanopyramid structures formed onto the OrmoStamp resist-coated glass substrate as the result of second imprint process with a scanned area of 5 × 5 μm. It can be seen that the inverted nanopyramid structures with periodic features and smooth surfaces have been precisely reproduced onto the OrmoStamp resist-coated glass substrate with high fidelity. **Figure 22(c)** illustrates the AFM image cross-sectional traces of the replicated inverted nanopyramid structures on resist-coated substrate. It shows that the replication from the upright nanopyramid on the resist-coated substrate is very similar and high accuracy compared to original Si master mold as illustrated in **Figure 18(c)**.

There are some issues which might come up during the development of the UV-NIL process. The adhesions between the resist and substrate are critical issues when no adhesion promoter is added to the substrate before the resist spin-coating process. In order to resolve the adhesion problem between the resist and the substrate, OrmoPrime08 from micro resist technology is used to enhance the adhesion of the OrmoStamp resist to the substrate. It is observed that the F_{13}-TCS SAM-coated master mold can be used several times in the imprint process, whereas imprint process will not be successful without F_{13}-TCS SAM coating due to the sticking and

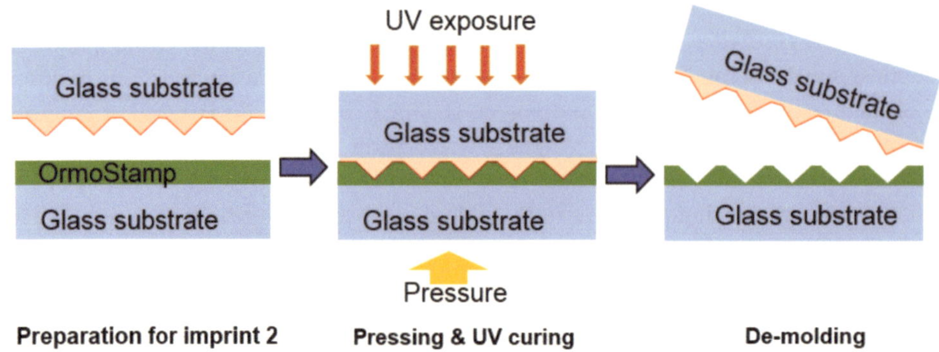

Figure 21. The schematic diagram of the overall process steps to replicate the inverted nanopyramid structures from upright pyramids mold.

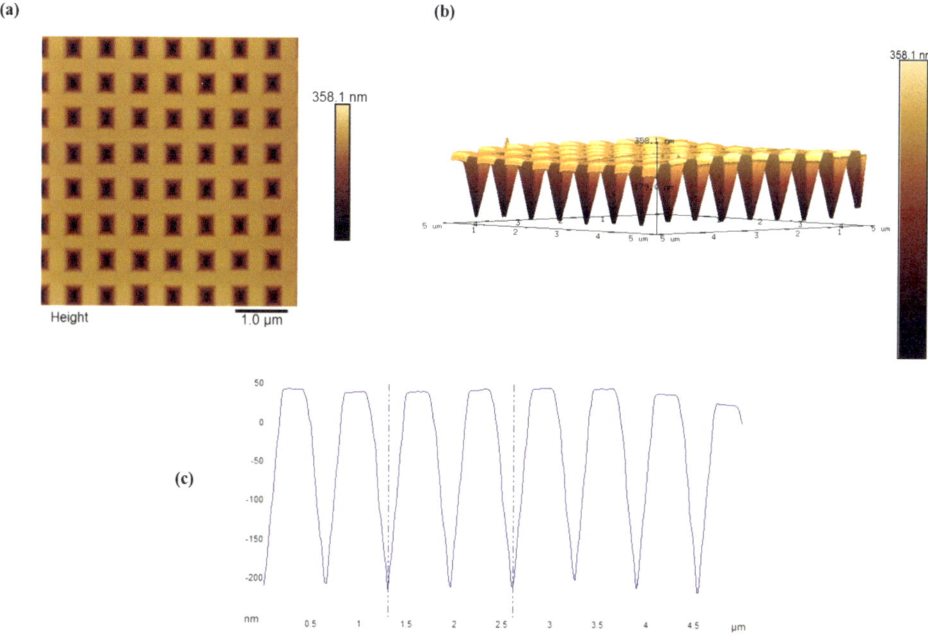

Figure 22. AFM images of inverted nanopyramid structures formed on UV transparent OrmoStamp resist-coated glass substrate after the second imprint: (a) 2-D view, (b) 3-D view, and (c) cross-sectional traces.

particles contamination. SAM coating enhances the imprint qualities and also improves the mold lifetime remarkably by precluding surface damage and contamination. It is also observed that the adhesion between the mold and the replica is stronger when higher exposure dose is used during the imprint process. This effect may be correlated to higher shrinkage of the resist when exposed to higher UV exposure dose. In this case, it is preferred to maintain constant UV exposure dose (1000 $mJcm^{-2}$) prior to the demolding, and additional UV exposure and hard baking must be done after demolding.

The inverted nanopyramid structures, which are utilized for light trapping in solar cells, can be transferred to the resist-coated substrate by the nanoimprint process without any structural losses. The inverted nanopyramid structures replicated by nanoimprint process can be used in different configurations (upright, inverted, within, or on top of substrates) to enhance the solar cell performance or antireflection coatings on glass are used.

4. Conclusions

In this chapter, the fabrication and replication of periodic nanopyramid structures by LIL and UV-NIL are presented. The inverted nanopyramid structures were fabricated on the Si substrates by LIL and subsequent pattern transfer process using reactive ion etching, followed by KOH wet etching. The pattern of nanoholes is recorded on AZMiR 701 i-line positive photoresist using LIL by double exposure. The CHF_3/Ar plasma etching is performed to transfer the nanoholes pattern into thin SiO_2 interlayer. Then, O_2 plasma etching is performed to transfer the pattern into ARC layer with thin SiO_2 as a mask. The pattern is transferred into the thermal oxide layer using CHF_3/Ar plasma etching. The inverted pyramid structures are formed on Si substrate by KOH wet etching, and the SiO_2 mask layer is removed by buffered HF etching. The fabricated inverted nanopyramid structures on Si substrate are utilized as a master mold in the nanoimprint replication process.

In the first nanoimprint process, the upright nanopyramid structures are formed on the OrmoStamp-coated glass substrate using Si master mold with high fidelity. The upright nanopyramid-structured glass substrate could be used as cover glass for solar cell application and as a mold for the second imprint process. In the second nanoimprint process, the inverted nanopyramid structures are fabricated on the OrmoStamp-coated substrate using the upright nanopyramid-structured glass substrate as a mold. The replicated inverted nanopyramid structure on a resist-coated substrate is faithfully resolved with the high accuracy compared to original Si master mold. The upright and inverted nanopyramid structures by the nanoimprint process can be utilized as light-trapping and self-cleaning surfaces for different types of solar cells.

Author details

Amalraj Peter Amalathas* and Maan M. Alkaisi

*Address all correspondence to: amalraj.peteramalathas@pg.canterbury.ac.nz

Department of Electrical and Computer Engineering, MacDiarmid Institute for Advanced Materials and Nanotechnology, University of Canterbury, Christchurch, New Zealand

References

[1] Lee SH, Han KS, Shin JH, Hwang SY, Lee H. Fabrication of highly transparent self-cleaning protection films for photovoltaic systems. Progress in Photovoltaics: Research and Applications. 2013;**21**:1056-1062. DOI: 10.1002/pip.2203

[2] Sivasubramaniam S, Alkaisi MM. Inverted nanopyramid texturing for silicon solar cells using interference lithography. Microelectronic Engineering. 2014;**119**:146-150. DOI: 10.1016/j.mee.2014.04.004

[3] Amalathas AP and Alkaisi MM. Enhancing the performance of solar cells with inverted nanopyramid structures fabricated by UV nanoimprint lithography. 2016 IEEE 43rd Photovoltaic Specialists Conference (PVSC); 2016. IEEE

[4] Amalathas AP, Alkaisi MM. Efficient light trapping nanopyramid structures for solar cells patterned using UV nanoimprint lithography. Materials Science in Semiconductor Processing. 2017;**57**:54-58. DOI: 10.1016/j.mssp.2016.09.032

[5] Kanamori Y, Sasaki M, Hane K. Broadband antireflection gratings fabricated upon silicon substrates. Optics Letters. 1999;**24**:1422-1424. DOI: 10.1364/Ol.24.001422

[6] Amalathas AP, Alkaisi MM. Periodic upright nanopyramid fabricated by ultraviolet curable nanoimprint lithography for thin film solar cells. International Journal of Nanotechnology. 2017;**14**:3-14. DOI: 10.1504/Ijnt.2017.082435

[7] Tong HD, Jansen HV, Gadgil VJ, Bostan CG, Berenschot E, van Rijn CJ, Elwenspoek M. Silicon nitride nanosieve membrane. Nano Letters. 2004;**4**:283-287. DOI: 10.1021/Nl0350175

[8] Hulteen JC, Van Duyne RP. Nanosphere lithography: A materials general fabrication process for periodic particle array surfaces. Journal of Vacuum Science & Technology A. 1995;**13**:1553-1558. DOI: 10.1116/1.579726

[9] Cheng JY, Ross CA, Thomas EL, Smith HI, Vancso GJ. Fabrication of nanostructures with long-range order using block copolymer lithography. Applied Physics Letters. 2002;**81**: 3657-3659. DOI: 10.1063/1.1519356

[10] Low HY. Complex and useful polymer micro-and nanostructures via nanoimprint lithography. International Journal of Nanotechnology. 2007;**4**:389-403. DOI: 10.1504/Ijnt.2007.013973

[11] Amalathas AP, Alkaisi MM. Upright nanopyramid structured cover glass with light harvesting and self-cleaning effects for solar cell applications. Journal of Physics D: Applied Physics. 2016;**49**:465601. DOI: 10.1088/0022-3727/49/46/465601

[12] McAlpine MC, Friedman RS, Lieber CM. Nanoimprint lithography for hybrid plastic electronics. Nano Letters. 2003;**3**:443-445. DOI: 10.1021/nl034031e

[13] Pisignano D, Persano L, Mele E, Visconti P, Anni M, Gigli G, Cingolani R, Favaretto L, Barbarella G. First-order imprinted organic distributed feedback lasers. Synthetic metals. 2005;**153**:237-240. DOI: 10.1016/j.synthmet.2005.07.273

[14] Cheng X, Hong Y, Kanicki J, Guo LJ. High-resolution organic polymer light-emitting pixels fabricated by imprinting technique. Journal of Vacuum Science & Technology B. 2002;**20**:2877-2880. DOI: 10.1116/1.1515307

[15] Zhang W, Chou SY. Fabrication of 60-nm transistors on 4-in. Wafer using nanoimprint at all lithography levels. Applied Physics Letters. 2003;**83**:1632-1634. DOI: 10.1063/1.1600505

[16] Falconnet D, Pasqui D, Park S, Eckert R, Schift H, Gobrecht J, Barbucci R, Textor M. A novel approach to produce protein nanopatterns by combining nanoimprint lithography and molecular self-assembly. Nano Letters. 2004;**4**:1909-1914. DOI: 10.1021/nl0489438

[17] Martın J, Nogues J, Liu K, Vicent J, Schuller IK. Ordered magnetic nanostructures: Fabrication and properties. Journal of Magnetism and Magnetic Materials. 2003;**256**:449-501. DOI: 10.1016/S0304-8853(02)00898-3

[18] Ahn S-W, Lee K-D, Kim J-S, Kim SH, Park J-D, Lee S-H, Yoon P-W. Fabrication of a 50 nm half-pitch wire grid polarizer using nanoimprint lithography. Nanotechnology. 2005;**16**: 1874. DOI: 10.1088/0957-4484/16/9/076

[19] Cao H, Tegenfeldt JO, Austin RH, Chou SY. Gradient nanostructures for interfacing microfluidics and nanofluidics. Applied Physics Letters. 2002;**81**:3058-3060. DOI: 10.1063/1.1515115

[20] Battaglia C, Escarré J, Söderström K, Charrière M, Despeisse M, Haug F-J, Ballif C. Nanomoulding of transparent zinc oxide electrodes for efficient light trapping in solar cells. Nature Photonics. 2011;**5**:535-538. DOI: 10.1038/Nphoton.2011.198

[21] Jošt M, Albrecht S, Kegelmann L, Wolff CM, Lang F, Lipovšek B, Krč J, Korte L, Neher D, Rech B. Efficient light management by textured nanoimprinted layers for perovskite solar cells. ACS Photonics. 2017;**4**:1232-1239. DOI: 10.1021/acsphotonics.7b00138

[22] Zhang C, Song Y, Wang M, Yin M, Zhu X, Tian L, Wang H, Chen X, Fan Z, Lu L. Efficient and flexible thin film amorphous silicon solar cells on nanotextured polymer substrate using sol–gel based nanoimprinting method. Advanced Functional Materials. 2017;**27**:1-8. DOI: 10.1002/Adfm.201604720

[23] Farhoud M, Hwang M, Smith HI, Schattenburg M, Bae J, Youcef-Toumi K, Ross C. Fabrication of large area nanostructured magnets by interferometric lithography. IEEE Transactions on Magnetics. 1998;**34**:1087-1089. DOI: 10.1109/20.706365

[24] Xie Q, Hong M, Tan H, Chen G, Shi L, Chong T. Fabrication of nanostructures with laser interference lithography. Journal of Alloys and Compounds. 2008;**449**:261-264. DOI: 10.1016/j.jallcom.2006.02.115

[25] Chen CG, Konkola PT, Heilmann RK, Pati G, Schattenburg ML. Image metrology and system controls for scanning beam interference lithography. Journal of Vacuum Science & Technology B. 2001;**19**:2335-2341. DOI: 10.1116/1.1409379

[26] Walsh ME. On the Design of Lithographic Interferometers and Their Application. Massachusetts Institute of Technology; 2004

[27] Ji R. Templated Fabrication of Periodic Nanostructures Based on Laser Interference Lithography [Ph.D. Dissertation]. Martin-Luther-Universität Halle-Wittenberg; 2008

[28] Park S, Cho E, Song D, Conibeer G, Green MA. N-type silicon quantum dots and p-type crystalline silicon heteroface solar cells. Solar Energy Materials and Solar Cells. 2009;**93**:684-690. DOI: 10.1016/j.solmat.2008.09.032

[29] Seidel H, Csepregi L, Heuberger A, Baumgärtel H. Anisotropic etching of crystalline silicon in alkaline solutions. I. Orientation dependence and behavior of passivation layers. Journal of the Electrochemical Society. 1990;**137**:3612-3626. DOI: 10.1149/1.2086277

[30] Chou SY, Krauss PR, Renstrom PJ. Imprint of sub-25 nm vias and trenches in polymers. Applied Physics Letters. 1995;**67**:3114-3116. DOI: 10.1063/1.114851

[31] Ahn SH, Guo LJ. High-speed roll-to-roll nanoimprint lithography on flexible plastic substrates. Advanced Materials. 2008;**20**:2044-2049. DOI: 10.1002/adma.200702650

[32] Colburn M, Johnson SC, Stewart MD, Damle S, Bailey TC, Choi B, Wedlake M, Michaelson TB, Sreenivasan S, Ekerdt JG. Step and flash imprint lithography: A new

approach to high-resolution patterning. In: Microlithography'99. International Society for Optics and Photonics; 1999

[33] Chou SY, Keimel C, Gu J. Ultrafast and direct imprint of nanostructures in silicon. Nature. 2002;**417**:835-837. DOI: 10.1038/nature00792

[34] Ruiz SA, Chen CS. Microcontact printing: A tool to pattern. Soft Matter. 2007;**3**:168-177. DOI: 10.1039/b613349e

[35] Bao L-R, Cheng X, Huang X, Guo L, Pang S, Yee A. Nanoimprinting over topography and multilayer three-dimensional printing. Journal of Vacuum Science & Technology B. 2002;**20**:2881-2886. DOI: 10.1116/1.1526355

[36] Haatainen T, Majander P, Riekkinen T, Ahopelto J. Nickel stamp fabrication using step & stamp imprint lithography. Microelectronic Engineering. 2006;**83**:948-950. DOI: 10.1016/j.mee.2006.01.038

[37] Vogler M, Wiedenberg S, Mühlberger M, Bergmair I, Glinsner T, Schmidt H, Kley E-B, Grützner G. Development of a novel, low-viscosity UV-curable polymer system for UV-nanoimprint lithography. Microelectronic Engineering. 2007;**84**:984-988. DOI: 10.1016/j.mee.2007.01.184

[38] Plachetka U, Bender M, Fuchs A, Vratzov B, Glinsner T, Lindner F, Kurz H. Wafer scale patterning by soft UV-nanoimprint lithography. Microelectronic Engineering. 2004;**73**:167-171. DOI: 10.1016/j.mee.2004.02.035

[39] Lee J, Park S, Choi K, Kim G. Nano-scale patterning using the roll typed UV-nanoimprint lithography tool. Microelectronic Engineering. 2008;**85**:861-865. DOI: 10.1016/j.mee.2007.12.059

[40] Fuchs A, Bender M, Plachetka U, Kock L, Koo N, Wahlbrink T, Kurz H. Lithography potentials of UV-nanoimprint. Current Applied Physics. 2008;**8**:669-674. DOI: 10.1016/j.cap.2007.04.019

[41] Beck M, Graczyk M, Maximov I, Sarwe E-L, Ling T, Keil M, Montelius L. Improving stamps for 10 nm level wafer scale nanoimprint lithography. Microelectronic Engineering. 2002;**61**:441-448. DOI: 10.1016/S0167-9317(02)00464-1

[42] Jaszewski R, Schift H, Schnyder B, Schneuwly A, Gröning P. The deposition of anti-adhesive ultra-thin Teflon-like films and their interaction with polymers during hot embossing. Applied Surface Science. 1999;**143**:301-308. DOI: 10.1016/S0169-4332(99)00014-8

[43] Jang E-J, Park Y-B, Lee H-J, Choi D-G, Jeong J-H, Lee E-S, Hyun S. Effect of surface treatments on interfacial adhesion energy between UV-curable resist and glass wafer. International Journal of Adhesion and Adhesives. 2009;**29**:662-669. DOI: 10.1016/j.ijadhadh.2009.02.006

[44] Escarre J, Söderström K, Battaglia C, Haug F-J, Ballif C. High fidelity transfer of nanometric random textures by UV embossing for thin film solar cells applications. Solar Energy Materials and Solar Cells. 2011;**95**:881-886. DOI: 10.1016/j.solmat.2010.11.010

Chapter 3

Large-Area Nanoimprint Lithography and Applications

Hongbo Lan

Additional information is available at the end of the chapter

http://dx.doi.org/10.5772/intechopen.72860

Abstract

Large-area nanoimprint lithography (NIL) has been regarded as one of the most promising micro/nano-manufacturing technologies for mass production of large-area micro/nanoscale patterns and complex 3D structures and high aspect ratio features with low cost, high throughput, and high resolution. That opens the door and paves the way for many commercial applications not previously conceptualized or economically feasible. Great progresses in large-area nanoimprint lithography have been achieved in recent years. This chapter mainly presents a comprehensive review of recent advances in large-area NIL processes. Some promising solutions of large-area NIL and emerging methods, which can implement mass production of micro-and nanostructures over large areas on various substrates or surfaces, are described in detail. Moreover, numerous industrial-level applications and innovative products based on large-area NIL are also demonstrated. Finally, prospects, challenges, and future directions for industrial scale large-area NIL are addressed. An infrastructure of large-area nanoimprint lithography is proposed. In addition, some recent progresses and research activities in large-area NIL suitable for high volume manufacturing environments from our Labs are also introduced. This chapter may provide a reference and direction for the further explorations and studies of large-area micro/nanopatterning technologies.

Keywords: large-area nanoimprint lithography, large-area micro/nanopatterning, full wafer NIL, roller-type NIL, roll-to-plate NIL, roll-to-roll NIL

1. Introduction

Mass producing nanostructure over large areas is critical to the commercial applications of nanotechnology. Large-area nanopatterning has demonstrated great potential which can significantly enhance the performance of many devices and create innovative products, such as LEDs, solar cells, hard disk drives, laser diodes, displays, sub-wavelength optical elements, anti-reflective glass with moth's eye structures, flexible electronics, OLED, etc. [1–7]. For example, the

solar cells with submicro anti-reflective coating exhibited higher photocurrent and higher power conversion efficiency compared to those without nanostructures [8]. Nano-patterned sapphire substrates (NPSS) and photonic crystals (PhC) have been considered as the most effective approaches to improve the light output efficiency (internal quantum efficiency and external quantum efficiency) of LEDs and beam shaping [9, 10]. However, mass producing large-area nanostructures, particularly nanopatterning on various curved or non-flat surfaces or fragile substrates or flexible substrates, are particularly difficult using current patterning methods. Furthermore, a variety of existing micro/nano-manufacturing technologies such as optical lithography, electron beam lithography, focused ion beam lithography, interference lithography, etc., cannot cope with all the practical requirements of industrial scale applications with respect to high resolution, high throughput, low cost, large-area, especially patterning on non-flat and curved surface. It is now still a challenging issue to produce large-area micro/nanoscale structures with low cost, high throughput for existing micro/nano-manufacturing technologies [11].

Nanoimprint lithography (NIL) has now been regarded as a promising nanopatterning approach with low cost, high throughput, and high resolution, especially for fabricating the large-area micro/nanoscale patterns and complex 3D structures and high aspect ratio features. Due to these outstanding strengths and unique capabilities, it was accepted by International Technology Roadmap for Semiconductors (ITRS) in 2009 for 16 and 11 nm nodes, scheduled for industrial manufacturing in 2013. Toshiba has validated NIL for 22 nm and beyond. NIL has been listed as one of 10 emerging technologies that will strongly impact the world by MIT's Technology Review. The resolution potential has been demonstrated by the replication of 2.4-nm features. It is expected to play a critical role in the commercialization of nanostructure applications [12–14]. In particular, a variety of emerging large-area NIL processes (e.g. wafer-scale soft UV-NIL process, roll-to-plate NIL, roll-to-roll NIL process) provide ideal solutions and powerful tools for mass producing micro/nanostructures over large areas and continuous patterning at low cost and high yield rate for the industrial scale applications in compound semiconductor optoelectronics, wafer-level optical element, and nanophotonic devices, especially for LED patterning using full wafer NIL and anti-reflective films or coatings by roll-to-roll nanoimprint process. That opens the door and paves the way for many commercial applications not previously conceptualized or economically feasible.

A large number of studies and efforts regarding large-area nanoimprint lithography have been carried out by both academia and industries. Great progresses in large-area nanoimprint lithography have been achieved in recent years. This chapter focuses on the significant progresses in large-area NIL processes and industrial applications. The rest of the chapter is organized as follows. Typical and emerging large-area NIL processes are discussed in detail in Section 4. Furthermore, Section 3 presents some industrial applications of large-area NIL. Prospects, challenges, and future directions for industrial scale large-area NIL are elaborated in Section 5. Finally, Section 6 summaries this chapter.

2. Large-area NIL processes

A variety of large-area NIL processes have been proposed and developed by both academia and industries. According to the types of used molds, type of imprint contact, and imprinting continuity, large-area NIL technology can be classified into two fundamental categories: the full wafer NIL and roller-type NIL. Roller-type NIL can be further divided into two subclasses: roll-to-plate (R2P) NIL and roll-to-roll (R2R) NIL. The R2R NIL process has shown a highly promising future to be implemented as a full-scale production process due to their high throughput and large-area patterning capability. The continuous roller-pressing process has been currently applied in many industrial fields [15–17].

2.1. Full wafer NIL (plate-to-plate type NIL, batch press type)

In full wafer NIL, a (flexible) flat mold/stamp is utilized to press into a resist layer on a rigid substrate, resulting in a full field contact. In order to pattern a whole wafer area without the high imprinting forces and demolding force, soft UV-NIL using a flexible (or soft) mold has been proven to be a very promising approach for producing large-area patterns up to wafer-level in the micrometer and nanometer scale. The process possesses the following outstanding strengths: (1) flexible mold using; (2) sequential imprinting; and (3) peel-off demolding. Substrate conformal imprint lithography (SCIL) developed by Philips Research and SUSS MicroTec, is a novel large-area wafer-scale nanoimprint method with nanoscale resolution. It bridges the gap between small rigid imprint stamp for best resolution and large-area soft imprint with limited resolution. NIL is mainly limited by practicable reasons in imprint area due to the waviness of the substrates. To implement full wafer NIL, the SCIL adopted the composite mold and the sequential imprinting method. The composite mold is composed of two rubber layers on a thin glass support. The in-plane stiffness of the stamp avoids pattern deformation over large areas, while out-of-plane flexibility allows conformal contact to underlying surface features. SCIL is based on a sequential imprinting principle, whereby the soft mold is placed gradually on the substrate and is then removed (see **Figure 1**). The wavelike progression of the contact front minimizes air inclusions even on large areas, and the sequential separation of the mold and substrate (peel-off demolding) allows for a clean and reliable disconnection that does not damage the patterned structures. Currently, the SCIL has achieved sub-10-nm resolution on wafer-scale areas (6 inch wafer). The SCIL technology can well cope with non-flat substrates and implement full wafer nanoimprinting in a single step. For the SCIL process, the

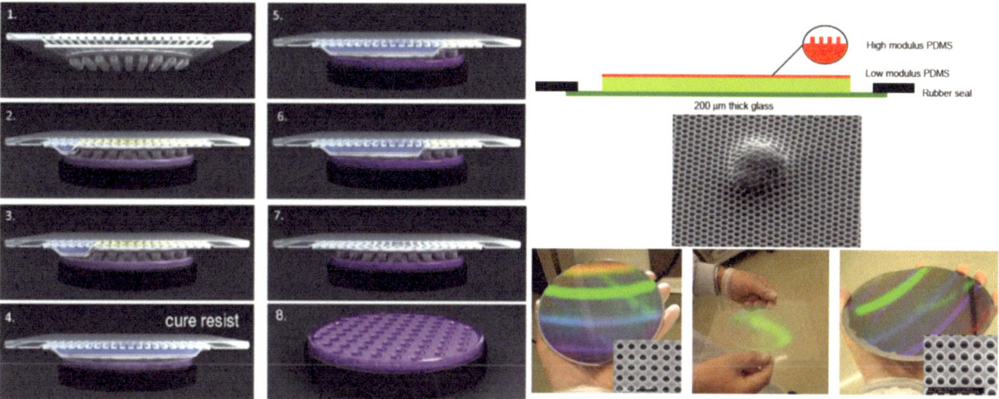

Figure 1. Schematic diagram of substrate conformal imprint lithography (SCIL), composite mold, and photo of imprinted patterns [1, 18].

curing of sol-gel resist relies on the diffusion of solvents into the PDMS stamp. Depending on the operation and preparation conditions, the curing time varies from 5 to 15 min. In order to shorten curing time of the process, a UV-enhanced SCIL process using UV curable material has been developed. Fader et al. introduced UV-SCIL with purely organic UV-curing materials showing curing times of 17 s [1, 18–19]. The SCIL technology can well cope with non-ideal substrates and implement full wafer imprinting in a single step. The excellent performance of SCIL in respect to substrate conformity and pattern fidelity over large areas enables this nanoimprint technology which is a powerful tool, especially for applications like LED/VCSEL, photonic crystal LEDs, three-dimensional (3D) photonic structures, nano-patterned sapphire substrates, optical elements, or patterned media [1].

Lan et al. reported a full wafer soft UV-NIL using a composite mold with tri-layer structure. The composite mold comprises of a thin layer of fluoropolymer-based material as the patterning layer, a thick layer of *s*-PMDS as intermediate flexible or cushion layer, and a thin PET (or glass sheet) as the support layer. **Figure 2** illustrated the schematic diagram of the proposed full wafer soft UV-NIL process. The imprinting process is performed by a sequential and micro-contacting solution starting from the center to two sides of the mold. The separation process employs a continuous 'peel-off' demolding mode starting from two sides to the center of the mold. Compared to the SCIL, the distinct advantages of the process include: (1) the imprinting and demolding procedure take the mold center as axis of symmetry, are carried out at the same time

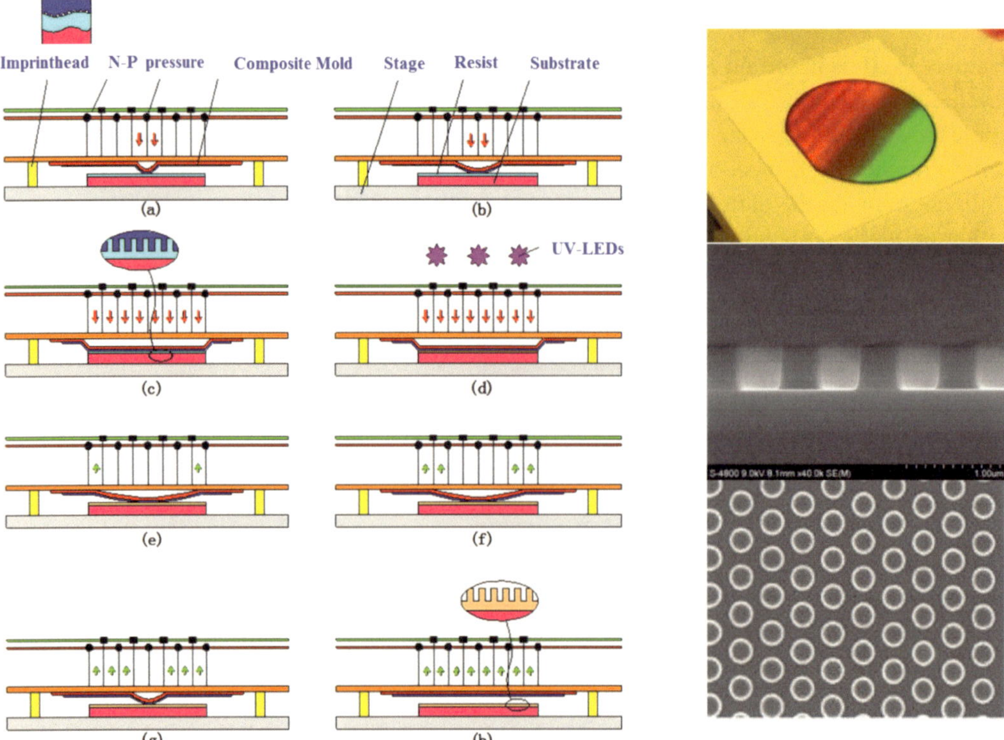

Figure 2. Schematic of full wafer nanoimprint lithography with a tri-layer composite mold and imprinted patterns (a-h).

on two sides with higher throughput and easier eliminating trapped air bubble; (2) an enhanced demolding approach is adopted; and (3) since the imprinting procedure is performed under a low vacuum pressure environment, it can better remove the trapped air bubbles and provide completely conformal contact [11].

Lan et al. proposed a new full wafer soft UV-NIL based on two air chambers. The imprinting and demolding processes are conducted by the close collaboration between the switching positive pressure and negative pressure in two air chambers and the upward and downward movement of substrate stage. **Figure 3** illustrated the basic principle and work flowchart of the proposed full wafer UV-NIL process and imprinted patterns. Compared to other full wafer soft UV-NIL, it has several outstanding advantages such as more compact configuration, easier operation, higher yield rate, easy to control, etc.

By integrating the imprint technology and roll-to-roll technology, which is based on extruder's coater technology, Toshiba Machine developed a nanoimprint equipment ST50S-LED for high-brightness LEDs. **Figure 4** shows the schematic diagram of the UV imprinting equipment. This imprinter enables UV imprint on the whole surface of a curved wafer in film mode, and has the features of high-precision control of the press position and press force. To support imprinting on a warped substrate or depositing defect substrate, a flexible resin mold is used in order to enable imprinting with high accuracy of form and less dispersion of residual resin layer. The rolled resin mold made by the roll-to-roll UV imprinting machine (CMT series) is used for automatic successive imprinting. A ST Head is employer to prevent un-uniform contact caused by parallelism difference between the substrate and mold to enable uniform transcription on the surface. This can enable automatic whole-surface UV imprint to form a fine-shape pattern of nanometer-order on a curved LED substrate. The maximum throughput of this equipment is 45 wafers/h [20, 21].

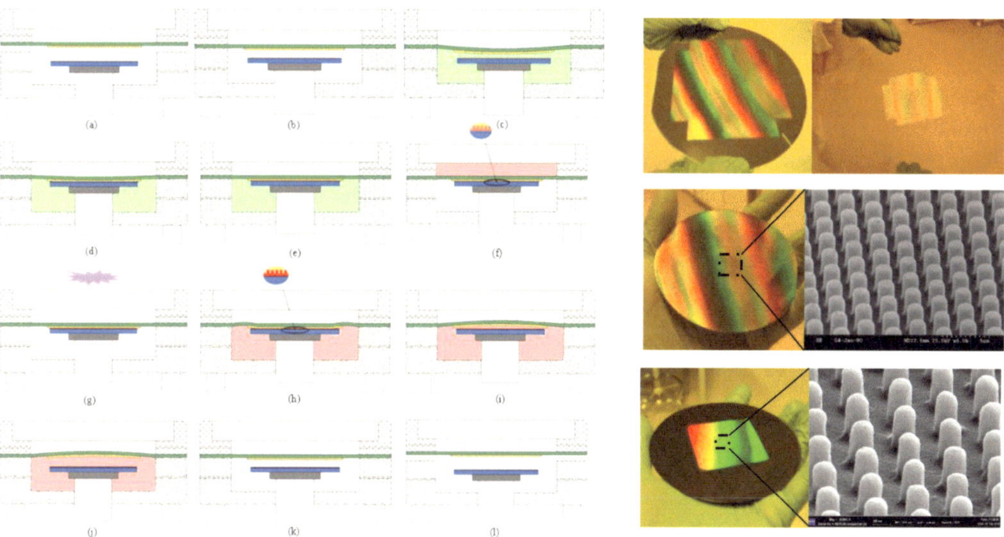

Figure 3. Schematic diagram of the principle of full wafer nanoimprint lithography with dual chambers, and imprinted patterns (a-l).

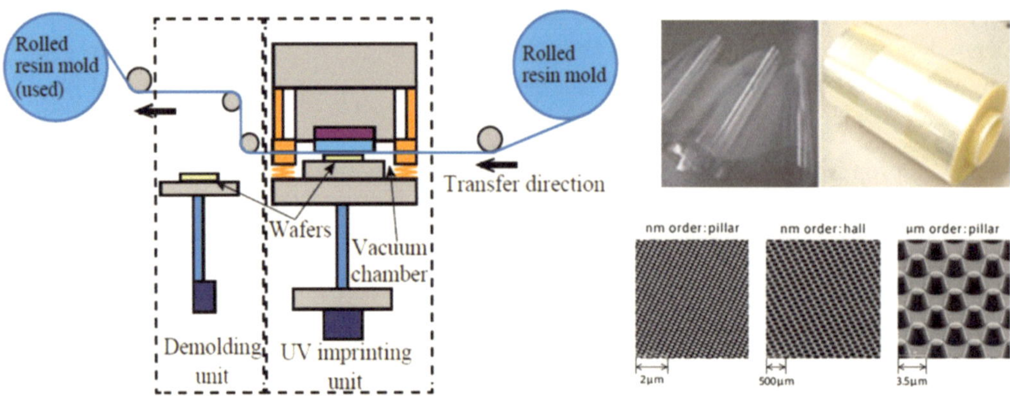

Figure 4. Schematic diagram of the UV imprinting equipment, resin mold (roll type) and imprinted patterns [20, 21].

A roller press mechanism is also used to in full wafer NIL process. Since a roller press mechanism is utilized, the actual contact area during imprinting is only a line along the roller in contact with the substrate rather than the entire mold area. Furthermore, the roller-based NIL process has the advantage of reduced issues regarding trapped air bubbles, thickness variation, and dust pollutants, which also greatly improve its replication uniformity. Youn et al. reported the prototype development of a roller-based imprint system and its application to large-area polymer replication for a microstructured optical device. **Figure 5** demonstrated its basic principle and process layout. A roller is utilized to press a flat flexible mold supported by several coil springs onto the polymer substrate. As the roller imprints onto the substrate via platform movement, pullers will be automatically elevated to lift and separate the flexible mold from the substrate. The system has the capacity to replicate ultra-precision structures on an area of 100 mm × 100 mm at the scanning speed range of 0.1–10 mm/s. Feature sizes down to 0.8–5 µm have been reported to be successfully imprinted [15, 22].

ESCO in Japan developed a Roller Press Scan® method (pressing sequentially using a roller), as shown in **Figure 6**. A roll is utilized to directly press the substrate that is placed onto resin-coated mold rather than the flexible mold. The maximum imprint area is 450 mm × 500 mm. The process has been applied to many fields such as moth-eye sheet, photonic crystal, optical waveguide (several tens of micrometers), micro-lens array (few millimeters), etc. [23].

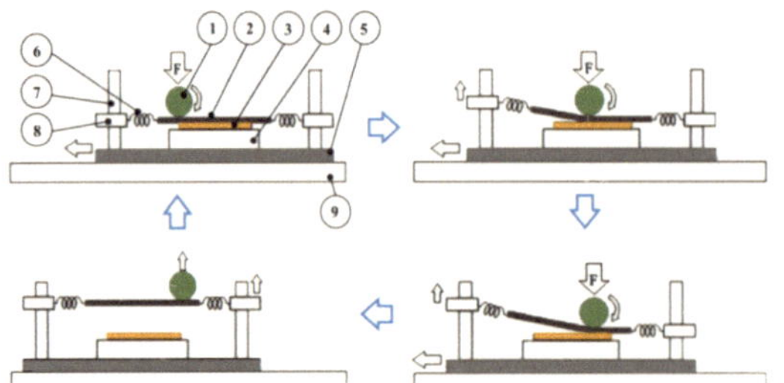

Figure 5. Process flow chart of the roller-based NIL process [22].

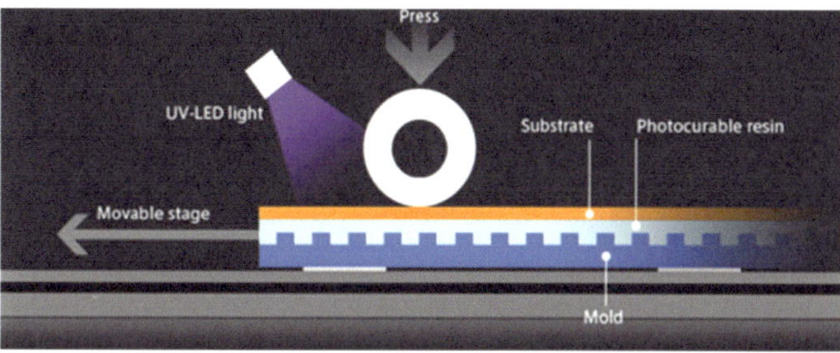

Figure 6. Schematic of the Roller Press Scan® method and imprinted patterns [23].

Soft UV-NIL has been considered as one of the most promising methods for mass producing micro- and nanoscale structures on the entire wafer at low cost for the applications in compound semiconductor optoelectronics and nanophotonic devices, especially for LED patterning.

Traditional plate-to-plate type NIL using a large mold is the simplest way for producing large-area nanostructures, but commonly requires high pressure that increases the possibility of mold deformation even the damage of the mold and the substrate. Further, the thickness variation on a mold or substrate can be as large as a micrometer on a large wafer area, and it is hard to be compensated. During imprinting, non-conformal contact occurred by the local flatness distortion in the mold causes the reduction of replication uniformity and leads to a huge stress concentration, resulting in the warping or distortion of the stamp. Therefore, soft UV-NIL by using a flexible mold has been proven to be a cost-effective high volume nanopatterning method for large-area structure replication up to wafer-level (up to 300 mm) in the micrometer and nanometer scale, fabricating complex 3D micro/nanostructures, especially making large-area patterns on the non-planar surfaces even curved substrates at low cost and with high throughput.

As a result, in order to better implement the full wafer NIL, some critical issues and promising solutions are described as follows. For the imprinting process, four important issues must be considered: (1) achieving uniform pressure distribution across the full wafer; (2) ensuring entirely conformal contact on the imprinting full field; (3) avoiding trapped air bubble defects; and (4) reducing imprinting forces. Applying gas-assisted press, sequential contact, vacuum imprint environment, and composite flexible mold may be effective solution. For the demolding process, peel-off demolding, fluoropolymer-based mold, and PVA-based thin-film mold can be used to avoid demolding defects and mold damage. For full wafer NIL, how to fabricate a large size wafer-level master has been became as the most challenging issue.

2.2. Roller-type NIL

In full wafer NIL (P2P), the entire imprint area is imprinted in a single imprinting cycle regardless of its size. However, this method is unsuitable for very large imprinting areas as it would require especially large imprint force, which may reach 20 kN of force for an 8-in. wafer [15]. The roller-type NIL (RIL) provides a unique solution to these challenging issues encountered in the conventional wafer-scale NIL process, because only a line area is in contact during imprinting, thus requiring a much smaller force to replicate the patterns. Moreover, because the stamp used in R2R NIL is in the form of a roller shape, the stamp-substrate separation proceeds in a "peeling-off" mode, which requires very small demolding force.

The roller-type NIL (RIL), first proposed by the Chou group [24], has now become one of the most promising candidates for rapid patterning on a very large-area substrate. Typical RNIL has a roller-shaped mold which contacts to a counter roller. A substrate goes through it and micro/nanostructures are transferred from the roller-shaped mold to the substrate. This technique can realize continuous imprinting, and can transfer patterns to the rigid or flexible substrates. Compared to P2P NIL, RNIL has the advantages of better uniformity, a lower imprint force, and the ability to repeat the patterning process continuously on a large substrate. Additionally, due to the line contact, the roller-based NIL process has the advantage of reduced issues regarding trapped air bubbles, thickness variation, and dust pollutants, which also greatly improve its replication uniformity.

2.2.1. Roll-to-plate type NIL

The roll-to-plate (R2P) NIL process uses a roller mold to pattern large-area rigid substrate. A roller mold is employed to imprint various rigid substrates such as glass substrates. The typical applications include flat panel displays, solar cells, anti-reflective glasses, etc. In 1998, Steven Chou's group firstly demonstrated a R2P thermal NIL process. A roller mold, which was made by wrapping a 100 μm thick nickel compact disk master mold with sub-micron features around the smooth roller, is imprinted and rolled on the substrate. Sub-100 nm resolution pattern transfer has been achieved using this method [24].

Ahn et al. described a R2P UV-NIL process for fabricating large-scale nano- and micropatterns on rigid substrates, as shown in **Figure 7**. The UV roll nanoimprinting system has the following components: a dispensing unit, a pair of flattening rollers, a UV light illumination unit, gap

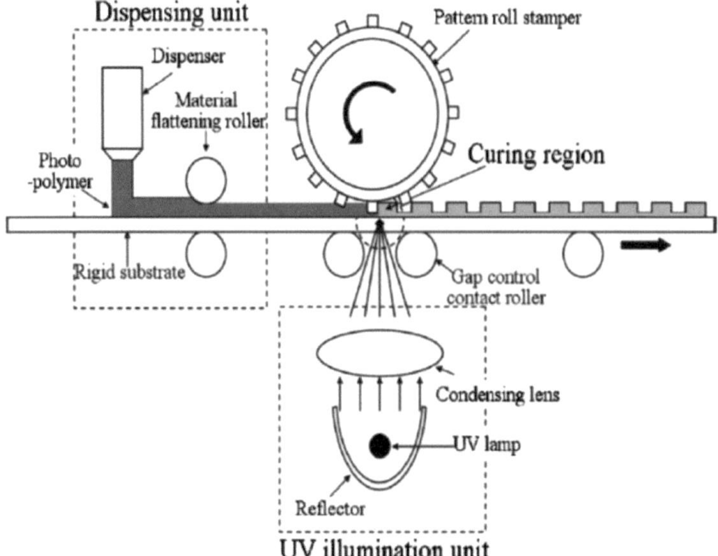

Figure 7. Schematic of R2P UV-NIL process for a rigid and transparent substrate developed by Ahn et al. [25].

control contact rollers, and a roller mold. The pair of flattening rollers enables the uniform coating of the photopolymer onto the substrate, which is achieved by controlling the size of the gap between the rollers. To obtain thickness uniformity, passive gap control by pressing the contact roller against the roll stamper was used. However, there is a limitation of the process since the substrates used should be transparent [25].

Park reported a R2P UV-NIL tool and process using double-layered soft cylindrical stamp to replicate nanoscale patterns on a Si substrate, as shown in **Figure 8**. 30 nm dot array patterns have been obtained. The R2P UV-NIL tool consists of a transparent quartz cylinder controlled by two synchronized motors, a passive compliant stage, a pressing unit in the Z direction, a UV system with a wavelength of 365 nm, and a fine stage with 3 nm resolution. Because of the synchronized motors used, there is no slip or misalignment between the replica and the quartz cylinder due to the tight fitting and compliant stage of the imprinting process. The trapped air problem is continually eliminated by means of the line of contact between the cylinder and the flat substrates from the first contact position to the last contact position. Ahn and Guo demonstrated large-area (4 in. wide) continuous imprinting of nanoscale structures by using a newly developed 6 in.-capable roll-to-plate (R2PNIL) apparatus. The grating patterns with 300 nm line width have been continuously transferred on glass substrates with greatly enhanced throughput [26].

Hitachi, Ltd. developed a sheet nanoimprint technology which enable 100 times higher productivity of nanoscale patterns, as shown in **Figure 9**. The enormous productivity is enabled by continuous processing of heating, pressing, cooling and separation using belt-shaped nanomold. A 200 nm in diameter and 240 nm tall dots (aspect ratio 1.4) were formed directly onto a 15 m-long polystyrene film. The process has some outstanding advantages such as extending cooling time, without size limitation of the mold [27].

In order to achieve transferred micropatterns with a precise profile yet no residual layer on the substrate, a roller-reversal imprint (RRI) process was proposed as shown in **Figure 10**. The resist is dispensed onto the roller mold using slot die instead of being coated onto the substrate, allowing it to fill in the mold cavities. A doctor blade is employed to remove excessive resist from the roller mold as it rotates. Upon contact with the substrate, the resist is transferred onto the substrate in a similar manner to a gravure printing. The transferred resist will then be

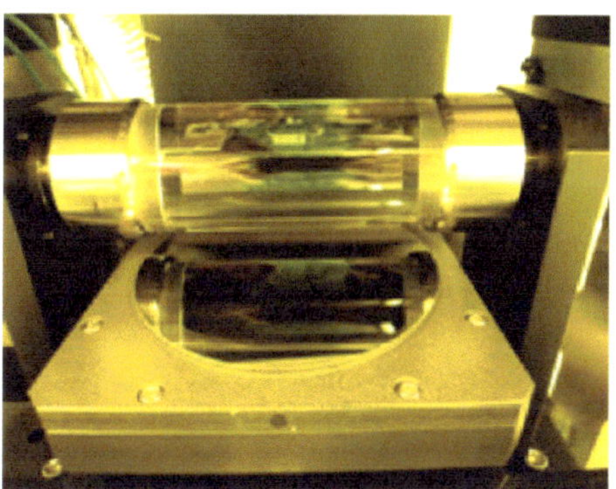

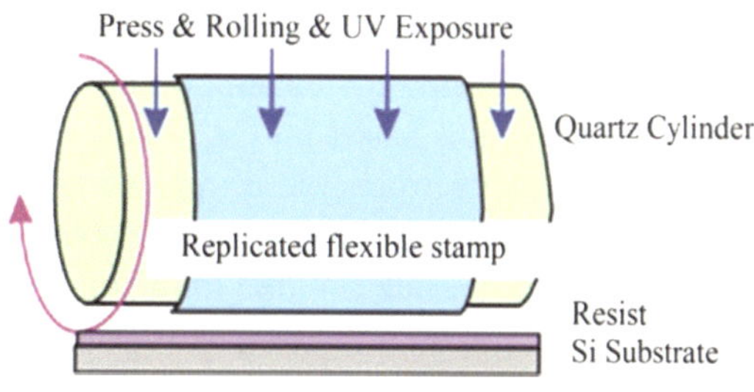

Figure 8. R2P UV-NIL using double-layered soft cylindrical stamp [26].

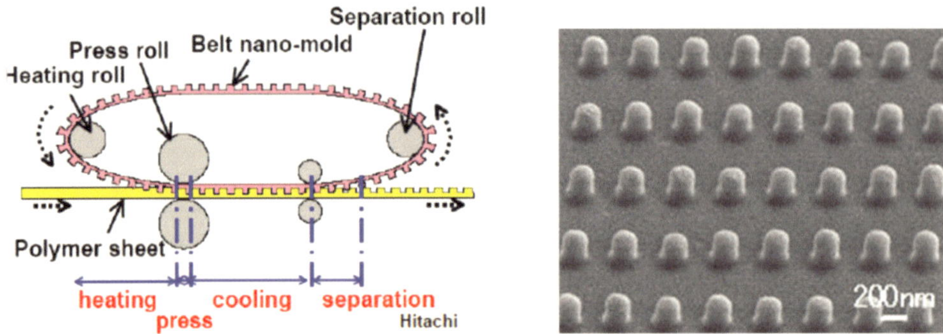

Figure 9. Schematic of sheet NIL developed by Hitachi and imprinted high aspect ratio pattern [27].

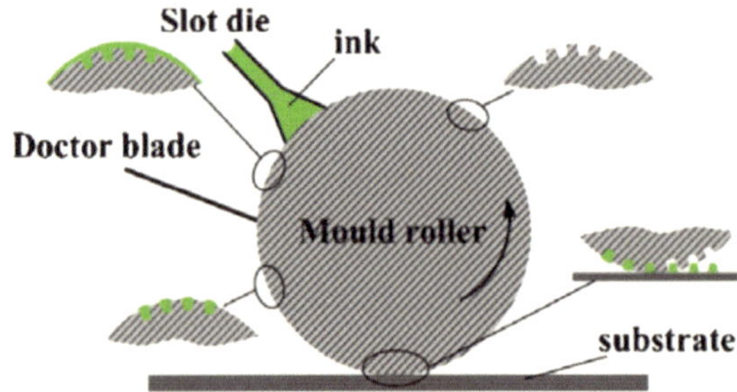

Figure 10. Schematic diagram of roller-reversal imprint process [28].

solidified by either UV or thermal curing. Feature sizes ranging from 20 to 130 μm in line width and 10–100 μm in depth have been successfully patterned using the roller-reversal imprint method [28].

While R2P NIL methods have great advantages over conventional P2P NIL in terms of imprint force, throughput, and size of equipment, it still has several limitations in realizing a continuous imprinting process.

2.2.2. *Roll-to-roll type NIL*

For the R2R NIL process, a roller mold is utilized to imprint onto a flexible substrate (or resist-coated flexible substrate) on a supporting roller instead of a flat plate in R2P NIL processes. The entire process is based on the roll-to-roll manufacturing concept, which has the advantages of continuous process, large-area patterning, and high throughput. It provides a highly promising solution for industrial scale applications. Therefore, the continuous R2R NIL technique offers a promising solution for high-speed large-area nanoscale patterning with greatly improved throughput. Moreover, it can overcome the challenges faced by conventional NIL in maintaining pressure uniformity and successful demolding in large-area imprinting [5, 15]. R2R NIL processes can be generally separated into two groups: thermoplastic roller imprinting and UV-curing roller imprinting.

2.2.2.1. R2R thermal NIL

Continuous imprinting of thermoplastics with a roller mold is a promising technique because of its simplicity in terms of tool setup and materials. All that is required for patterning is adequate heat and mechanical pressure, a suitable roller mold and a thermoplastic web that can be patterned directly; thus avoiding the need to deposit a patterning medium (resist) onto a substrate and all the issues arising from the additional processing steps. In addition, the technique is flexible in terms of materials and substrate selection. Various materials for the roller mold and a wide variety of thermoplastic materials can be used as either cast films on a substrate or as a stand-alone web [5]. In addition, functional materials such as semiconducting or light-emitting polymers can be directly patterned with this technique. In particular, with a roller mold seamless patterning is achievable and this is very useful for large-area low cost manufacturing applications such as anti-reflection coatings, micro-lens arrays and wire grid polarizers (WGPs).

Existing investigations on thermal roller imprinting revealed poor pattern transfer fidelity, especially for high aspect ratio features. The traditional R2R imprinting process suffers from the lack of an effective holding and cooling stage so that the adverse effects from the viscoelastic nature of polymers are not managed. To solve this problem and further improve the production rate, a new extrusion roller imprinting process with a variotherm belt mold was proposed, as shown in **Figure 11**, and its prototype was also developed at a laboratory scale. The major components of the R2R tool include an extruder, a belt mold, a roll-to-roll setup, and an induction heating unit. The extruded polymer film is imprinted between the belt mold and the pressure roller. Due to the variotherm capability, the imprinted film is effectively cooled before it is released from the belt mold. The experimental results demonstrated that a 30 μm sawtooth pattern can be faithfully transferred to extruded polyethylene film at take-up speeds higher than 10 m/min [29].

3D and multilayered nanostructures are becoming important with the technological advances in nanodevices such as nanoelectromechanical systems, nanofluidic devices, and nanophotonics. Nagato et al. from University of Tokyo proposed an iterative roller imprint method of multilayered nanostructures by combination of imprinting and bonding process. Thermoplastic polymer film is imprinted using a nanostructured mold and heated rollers (first layer). The next imprinted thin film is thermally bonded on the backside of the first layer using other rollers. By repeating these processes, a multilayered nanostructure was fabricated. By this method, multilayered

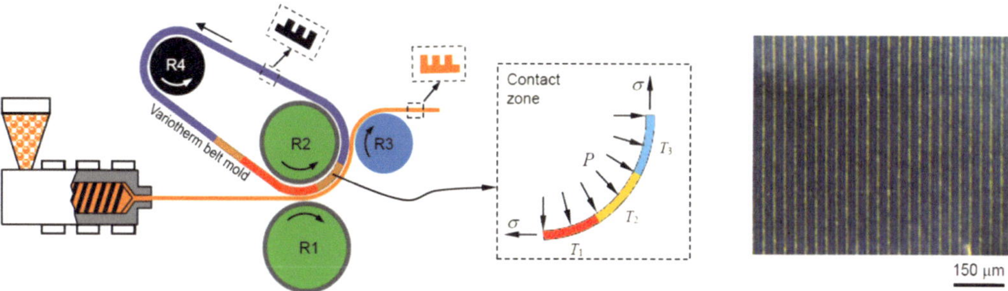

Figure 11. Schematic of extrusion roller imprinting process with a variotherm belt mold, imprinted structures [29].

nanostructures with a pitch of 800 nm and a depth of 300 nm were fabricated using PMMA. **Figure 12** illustrated the iterative roller imprint method of multilayered nanostructures and imprinted multilayered nanogaps with five-layered nanostructures [30].

In 2013, EVG releases the industry's first roll-to-roll thermal NIL tool (EVG®570R2R) which can mass produce films and surfaces with micro- and nanometer-scale structures for a variety of medical, consumer, and industrial applications, including microfluidics, plastic electronics, and photovoltaics. Recently, an EVG750® R2R hot embossing system was released, as shown in **Figure 13**. The EVG750R2R's innovative design provides excellent temperature and pressure uniformity for micro- and nanoscale patterning on a broad range of materials. It is designed to deliver the highest flexibility for R&D applications with a clear vision for automated mass manufacturing of flexible devices [31].

There are two challenging issues for R2R thermal NIL processes. One issue related to the imprinting of thermoplastic materials generally is the flow behavior in the vicinity of large or strongly nonuniform mold features. Another issue that must be solved is an effective cooling solution prior to demolding.

2.2.2.2. R2R UV-NIL

Although R2R thermal NIL has the advantage of simplicity and flexibility in terms of tool setup and materials, the process involves relatively high pressure (typically at least 5 MPa) and high temperature (typically about 100–300°C) due to the high viscosities of thermoplastic materials imprinted. Unlike with the imprinting of thermoplastic materials with R2R thermal NIL, R2R UV-NIL does not require elevated temperature or large applied pressures to create patterns which has shown higher throughput and lower product cost, because low pressures are used in UV roller imprinting, it is easier to protect the roller mold from accumulating defects due to particles and residues; thus improving mold lifetime and the overall quality of

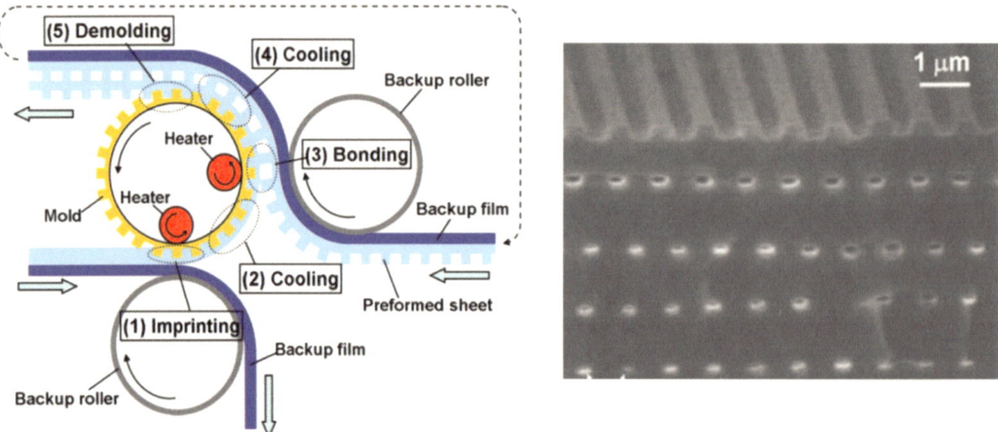

Figure 12. Schematic of roller imprint system for multilayered nanostructures, imprinted multilayered nanogaps with five-layered nanostructures [30].

Figure 13. Photo of the EVG750® R2R hot embossing system, imprinted, SEM image of 10 μm and 30 μm microfluidic structures replicated into PMMA, 500 nm holes replicated into PMMA [31].

replication [5]. In addition, low pressure requirements enable the use of soft mold materials such as ETFE, PDMS without risk of feature distortion.

Figure 14 illustrated a typical R2R UV-NIL system. It mainly consisted of the following functional components such as a dispensing unit, a doctor blade (or a pair of flattening rollers), a UV source unit, a roller mold, pressure rollers, a demolding roller, etc. [5]. A dispensing system is utilized to deposit a UV curable liquid resin either as a pattern of drops or as a continuous film. Following deposition of the resin, a variety of thickness control measures can be employed, such as a doctor blade. Multiple pressure rollers are often used to ensure

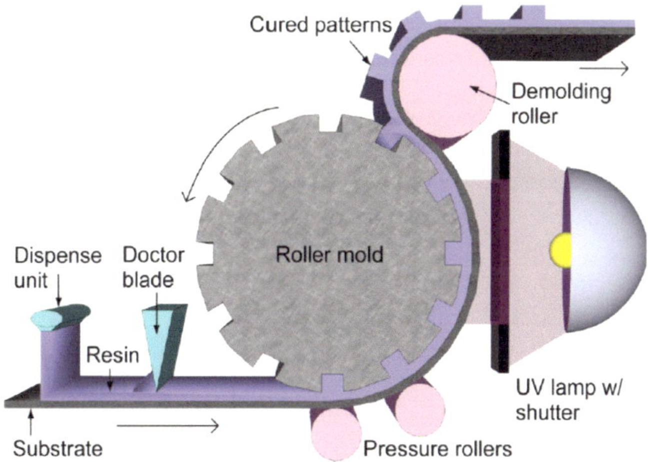

Figure 14. Schematic of a typical R2R UV-NIL system [5].

uniform spreading of the resin and filling of the roller mold cavities prior to UV exposure. A demolding roller is used to peel the cured patterns off the roller mold. How to extend roller mold lifetime is a challenging issue for the real commercialization or industrial applications of roller-type NIL. The roller mold should have good wear resistance properties. The mold material used should have low surface energy to ensure that the resist does not adhere to the mold surface during the demolding process, and to reduce friction during imprinting. A release coating on the roller mold surface can effectively prolong roller mold lifetime by preventing the adhesion of the resists. However, this release coating gradually deteriorates as the number of repetitions of NIL transfer increases. It is therefore important to maintain the lifetime of the release agent. A roller mold generally need be coated with a fluorinated silane anti-sticking layer to prevent sticking problems. In addition, mixing of various types of release agent is also an effective way in increasing the lifetime of the roller mold [15].

Guo et al. presented a R2R UV-NIL process for the fabrication of metal wire grid polarizers with nanoscale structures (down to 70 nm line width and 100 nm pitch) on a flexible substrate. **Figure 15** demonstrated the schematic of the R2R UV-NIL process for fabricating metal wire grid polarizer and imprinted patters. The system is composed of three functional modules: coating module, imprinting module, and metal deposition module. An ETFE flexible fluoropolymer stamp wrapped around a stainless steel roller is adopt to be the roller mold for the UV R2R NIL. The benefit of using ETFE is that it has good anti-stick properties (critical surface tension of 15.6 mN m^{-1}, cf. PDMS 19.6 mN m^{-1}) that do not deteriorate over many imprint cycles or from exposure to UV light, unlike fluorinated SAM anti-stick coatings [32].

A self-aligned imprint lithography (SAIL) was proposed by HP for solving alignment problem in R2R imprinting process, as shown in **Figure 16**. The SAIL has been used to fabricate precision electronic devices that require multiple alignment steps on a large dimensionally unstable substrate by incorporating a single imprinting step. The technique solves the problem of precision interlayer registry on a moving web by encoding all the geometry information required for the entire patterning steps into a monolithic three-dimensional imprint with

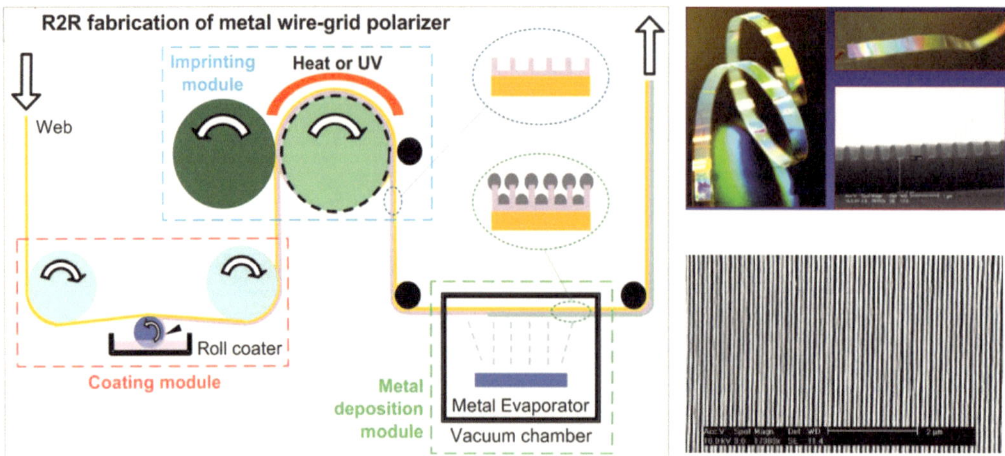

Figure 15. Schematic of R2R UV-NIL process for fabricating metal wire grid polarizer and imprinted patters [32].

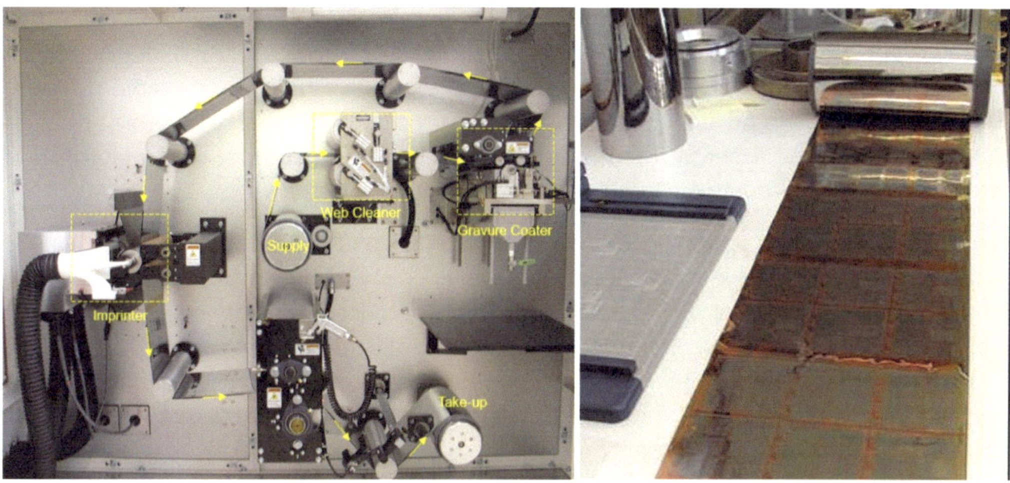

Figure 16. Photo of roll-to-roll coating and imprinting apparatus, imprinted active-matrix backplanes [33].

discrete thickness modulation. It can enable the patterning and alignment of submicron-sized features on meter-scaled flexible substrates in the R2R environment. The SAIL provides an effective way for mass producing flexible electronics and flexible display backplanes in particular for commercial production [33, 34].

Due to the limitations of current roller mold materials used for R2R NIL, John et al. described a large-area, continuous R2R nanoimprinting with PFPE composite molds. A R2R Nano Emboss 100 was also developed using the PFPE composite molds to address the challenge of fabricating nanostructured thin films on a high-speed, high-reliability platform. The tool consists of five sections, namely unwind, coating, imprinting, metrology/coating, and rewind, as shown in **Figure 17**. John et al. showed the efficiency, reliability, and durability of the PFPE-based

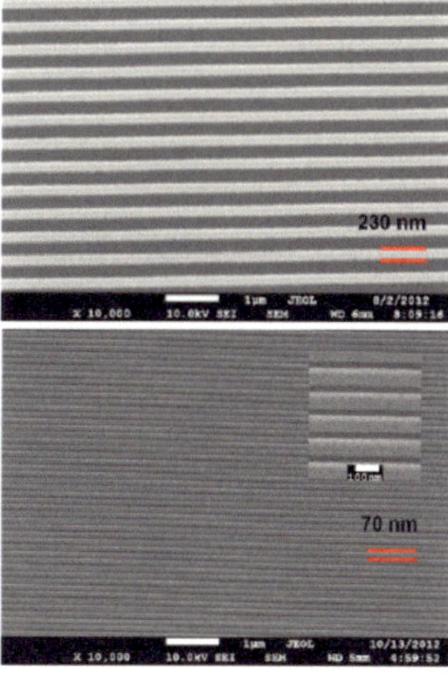

Figure 17. Photo of the custom designed R2R nanoimprinter; SEM images of 230 and 70 nm imprinted gratings on PET [35].

composite molds in replicating 1D and 2D micron to sub-100 nm features with high fidelity over successive imprinting cycles in a dynamic large-area high-speed roll-to-roll nanoimprinting process [35]. The experimental results have exhibited that the PFPE composite mold replicated nanofeatures with high fidelity and maintained superb mold performance in terms of dimensional integrity of the nanofeatures, nearly defect-free pattern transfer and exceptional mold recovering capability throughout hundreds of imprinting cycles [35].

A roll-to-roll type UV imprinting machine has been developed for mass producing resin molds, wire grid polarizers, flexible solar cells, etc. **Figure 18** illustrated the schematic of the R2R UV-NIL and imprinted products. An "intermittent coating" technology employing a gravure roll to mold UV curable resin coated on film is introduced. Sub-100 nm patterns, especially sub-30 nm patterns, onto ultraviolet (UV) curable resin on film substrate have been obtained. They also have investigated the potential of R2R UV-NIL for future wearable devices that would require large-area nanopatterning. This result indicates that the process has the potential to make sub-30 nm patterns on film [36].

Ahn and Guo illustrated large-area (4 in. wide) continuous imprinting of nanogratings by using a developed apparatus capable of roll-to-roll imprinting (R2R NIL) on flexible web and roll-to-plate imprinting (R2PNIL) on rigid substrate. For the proposed process, a belt-type mold is wrapped around two rolls. Either a flexible or rigid substrate coated with a resist mounted on another roll or a flat conveyer is then fed into the contact zone where a mold continuously imprints the replica pattern onto the substrate under conformal pressure [37]. The instant UV curing is followed at the outlet of the mold-substrate contact zone to finish the patterning. The 300 nm line width grating patterns can be continuously transferred on either glass substrate (roll-to-plate mode) or flexible plastic substrate (roll-to-roll mode) with greatly enhanced throughput. **Figure 19** showed the R2R NIL process and fabricated products [16, 35].

One significant advantage of the R2R NIL technique is that it inherited the high resolution pattern fidelity benefit from the traditional NIL technique with a drastically increased throughput. The possibility of defect generation during the mold-substrate separation process is considerably lower in the R2R NIL technique due to the peeling mode in which the substrate

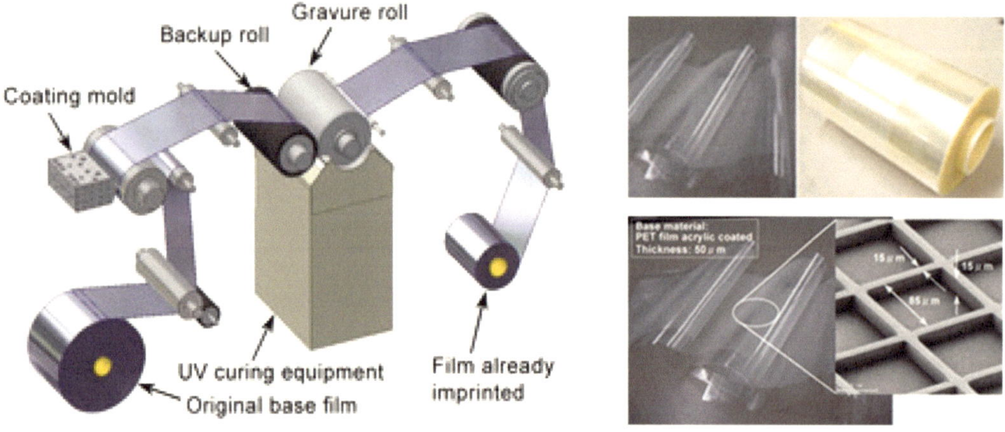

Figure 18. Schematic of R2R UV-NIL NIL developed by Toshiba, imprinted products (resin mold) [36].

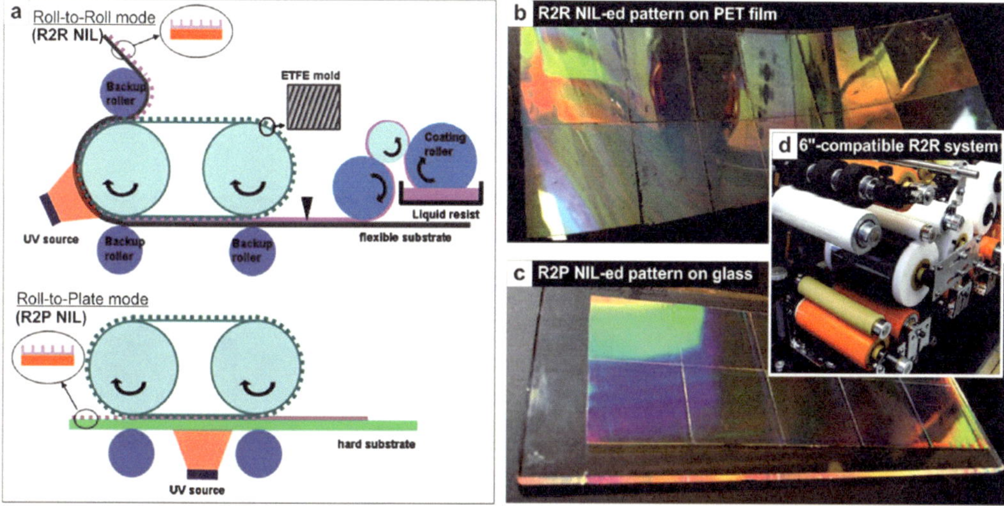

Figure 19. (a) Schematic descriptions of R2R NIL process (top) for flexible substrates and R2P NIL mode (bottom) for rigid substrates; (b) a PET film; and (c) a glass plate [35].

separates from the roller mold assembly as compared to the conventional NIL technique. In particular, R2R NIL has the unique capability of continuous patterning.

3. Applications of large-area NIL

Nanostructing has been one of the most promising and enabling technologies to enhance the performance of various products and devices ranging from high-brightness LEDs, high effective solar cells, high density storage to high clearness (HD) display. NIL has proven to be the most cost-efficient way to enable fabrication of nanopatterns on large areas as it is not limited by optical systems and can provide the best pattern fidelity for the smallest structures. Various types of large-area nanostructures and patterns including one-, two- and three-dimensional structures fabricated by NIL have been applied to diverse devices and products to enhance their performances. Large-area NIL technology paves the way for real-world nanotechnology products and innovative applications. Large-area NIL has now been utilized to fabricate various micro/nanostructures and devices for nanoelectronics, optoelectronics, nanophotonics, optical components, nanostructured glass, biological applications, etc. **Figure 20** demonstrated representative industrial applications and products with large-area NIL. Moreover, it has become a perfect match for some emerging application fields that are in great need of large-area patterning of submicro and nanoscale features at a low cost, such as patterned magnetic media, high-brightness light-emitting diodes (HB-LEDs), anti-reflective coatings or films, flexible electronics, printed electronics, OLED, wire grid polarizer, flat panel display, microfluidic devices, etc. In particular, this technique has demonstrated great commercial prospects in several market segments, HB-LEDs, anti-reflective coatings or films with moth-eye structures, flexible electronics, solar cells, architectural glass, WGP, optical elements, patterned media, micro-lens arrays, and functional polymer devices [1–5, 38–41].

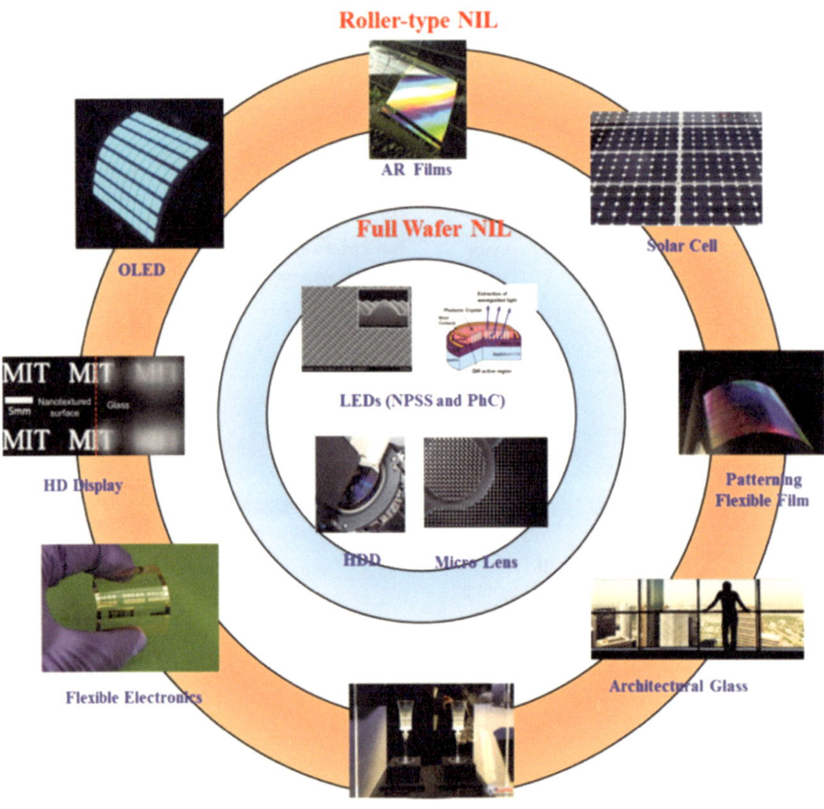

Figure 20. Typical industrial applications and products with large-area NIL.

3.1. Patterning LEDs

Light efficiency enhancement and manufacturing cost reduction have always been regarding as the two most crucial issues in LED industry, particularly for the large-scale realization of solid state lighting. Compared with other technologies improving the LED performance, two emerging techniques, photonic crystals and nano-patterned sapphire substrate (NPSS), have shown higher potential in output efficiency enhancement and beam shaping. NPSS and photonic crystal-based LEDs have become the most promising solutions for high-brightness LEDs (HB-LEDs). The typical characteristics of LED epitaxial wafers and sapphire substrates are with large variation in wafer topography (varying TTD), high bow and warp, surface roughness with surface protrusions with micron size and particle contaminations, etc. And these materials tend to be fragile or brittle. Due to the non-planar and rough nature of the LED epitaxial wafer and substrate, existing nanopatterning technologies cannot well meet the requirements of producing these nanostructures in both technology and cost level which mainly originated form the new challenging issues form LED patterning. Due to a very small depth of focus, optical lithography techniques have insufficiently fidelity for LED patterning. As warpage increases with larger wafer sizes, the ability of the photolithography tool to compensate for substrate warpage becomes even more critical. Interference lithography is another method of generating periodic patterns over large areas at low cost. Although the patterns made by IL are highly uniform and have superior long-range order, these patterns are usually in very simple geometric forms of grating lines and 2D dots, and their dimensions are difficult to reduce to sub-100 nm due to light diffraction. Furthermore, it is unsuitable for high volume production processes because the optical configuration has to be modified to realize different patterns. In addition, this approach requires a strict control of the environment to maintain stable fringe patterns. Full wafer NIL (soft UV-NIL with flexible mold) has the capability of nanopatterning on non-flat surface over large areas and is less-sensitive to the production atmosphere. Compared to ICs industry, the LED application is much more relaxed than IC's for overlay and defect density. Therefore, full wafer NIL has been considered as one of the most suitable solution for LED patterning. Due to its cost-effectiveness combined with superior processing performance, full wafer NIL will play a crucial role in moving the LED industry into a new realm of nanopattered LEDs with ultra-high efficiency. **Figure 21** showed some cases related to LED patterning using full wafer NIL. In addition, some commercial companies such as SUUS, Obducat, EVG, Toshiba, Aurotek, Luminus, etc. have been developing the process and equipment of full wafer NIL for high volume producing PhC LEDs and NPSS [17–20].

3.2. Photonics (anti-reflective coatings or films, wire grid polarizer)

Recent developments in many applications, such as photonics, HD displays, and flexible electronics, etc., have observed a rapid increase in demand for a lower cost, higher throughput and higher resolution micro/nanofabrication techniques. The continuous R2R NIL technique offers a promising solution for high-speed large-area nanoscale patterning with greatly improved throughput. In addition, the ability of micro- to nanometer-scale patterning on flexible substrates can enable many new applications in the area of photonics and organic electronics. Some typical applications include anti-reflective coatings and films, wire grid polarizers (WGPs), flexible electronics and flexible display backplanes, nanogratings, light enhancement coatings and films for displays, etc. Perhaps the application which is closest to mass production are anti-reflective coatings and films which are unique in that they are fabricated from homogenous sub-micron moth-eye structures such as cones or pillars. Because they are sub-micron structures and ideally composed of a single material, traditional roll-to-roll methods such as flexography or gravure printing cannot be employed. Continuous R2R UV-NIL is ideally suited, especially if the refractive index of the substrate web and the UV curable resin can be closely matched over the visible light spectrum. One of the most

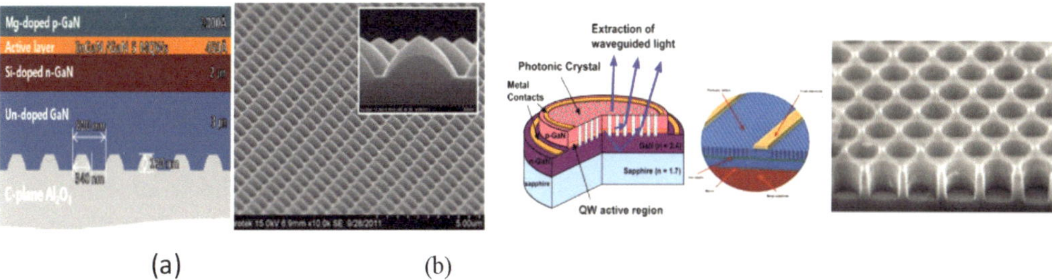

Figure 21. LED patterning using soft UV-NIL (a: NPSS-based LEDs; b: PhC LEDs) [19, 20].

promising applications for R2R thermal NIL is in the large-area fabrication of anti-reflective films, optical components and structures, particularly micro-lens structures.

LA-NIL has emerged as an effective approach to simultaneously control both the heterojunction morphology and polymer chains in organic photovoltaics. Currently, in the area of nanoimprinted polymer solar cells, much progress has been achieved in the fabrication of nanostructured morphology, control of molecular orientation/crystallinity, deposition of acceptor materials, patterned electrodes, understanding of structure-property correlations, and device performance [8]. Yang et al. reported low bandgap polymer solar cells with high efficiency of 5.5% can be achieved using NIL. This investigation indicates that solar cells made on the highest aspect ratio PCPDTBT nanostructures are among the best reported devices using the same material with a PCE of 5.5%, and NIL fabrication works better for low bandgap polymer solar cells [42]. Highly efficient colored perovskite solar cells having 10.12, 8.17, and 7.72% of the PCE for the red, green, and blue (RGB) colors, respectively, have been demonstrated by Lee et al. [43]. In this study, the large-area Ag nanogratings were fabricated by NIL-based processes.

In recent years, a number of commercial applications have been discovered which require low cost, large-area patterning, particularly displays, optical coatings and films, and biological applications such as anti-fouling surfaces and micro-fluidic devices. Wire grid polarizers (WGPs) are able to provide effective polarizer with high light transmittance and high contrast ratio. They are also thermally stable and structurally simple making them ideal for use as integrated polarizer in ultra-slim high performance display. Roller-type NIL has been utilized to mass produce the WGPSs on both the flexible and rigid substrates. A R2R UV-NIL process has been employed to fabricate the metal wire grid polarizers with nanoscale structures on a flexible substrate [4–8, 14–16].

4. Prospects and challenges of LA-NIL

Large-area NIL has been considered as a promising technology for cost-effective fabrication of sub-micron and nanopatterns over large areas. This technique has exhibited great potential and promising prospects in LEDs and HDD, and patterned substrates for full wafer imprinting, anti-reflective coatings or films with moth-eye structures, flexible electronics, and WGP for roll-to-roll imprinting, nanostructured glass for roll-to-plate type imprinting. Perhaps the application which is closest to mass production is anti-reflective coatings and films. LA-NIL has made great progress toward volume production.

Table 1 presented the summary and comparison of various large-area NIL processes. We can see that each process has its strength and weakness. According to the substrate type, imprint area, resolution, throughput, etc., it is necessary to determine the most suitable process for specific applications. Compared to the full wafer NIL, the roller-type NIL process possesses some distinctive advantages (lower imprint force and demolding force, better replication uniformity) because only a line area is in contact during imprinting. The line contact between the mold and the substrate during imprinting reduces the force for the complete filling, the effects of thickness unevenness and dust. Moreover, the roller-type NIL has more flexibility in the choice of replication geometry. In particularly, it has the unique capability of continuous patterning. R2R NIL process shows the highest yield rate.

Item	NIL process		
	Full wafer NIL (plate-to-plate type NIL)	Roller-type NIL	
		Roll-to-plate type NIL	Roll-to-roll type NIL
Substrate	Rigid substrate	Rigid substrate	Flexible substrate
Mold	Flat molds (rigid, flexible, or thin-film mold)	Roller mold or belt-style mold (flexible mold)	Roller mold (rigid or flexible mold)
Imprinting area	~300–400 mm^2 (450 mm × 500 mm)	1 m × 0.3 m (extremely large-area)	300 mm (width) (13 inch)
Resolution	Sub-20 nm	40 nm	Sub-30 nm
Continuous patterning	No	Yes	Yes
Throughput	Moderate (350 disks/h)	High	Very high (Web 30 m/min)
Industrial application	LED, HDD, wafer-level optical component	Solar cell panel, architectural windows, flat panel displays	Anti-reflective film and coating, flexible electronics, solar cells, displays, resin mold, WGP
Representative company or institute	SUSS, MII, Toshiba, Samsung, ESCO, AIST	Hitachi, Rolith	IMRE, EVG, Toshiba, HP, U Michigan, U Mass, U Tokyo
Distinctive features	Surface contact, non-flat substrate, peel demolding, gas-assisted or roller press force	Line contact, ultra-large-size rigid substrate, belt-type mold, extending cured time	Line contact, high throughput, large-scale production process, roller mold
Challenging issue	Large-area demolding, large size master fabrication	Belt-type mold fabrication, conformal contact	Seamless roller mold fabrication, coating method

Table 1. Summary and comparison of various large-area NIL processes.

Large-area imprinting technologies have made significant progresses. However, there are still many challenging issues, for example, defect, establishing an infrastructure, mold lifetime, large size master fabrication, uniformity of patterns and residual layer thickness across imprint field, discovering new applications and products especially suitable for large-area NIL, etc. Full wafer NIL is capable of wafer-scale processing at nanoscale resolution with relatively low cost. But the replica area is limited by the master mold size. It is difficult to ensure large-area conformal contact between the mold and substrate, especially for rigid stamp and non-flat substrate. For R2R UV-NIL process, it faces the following challenging issues: obtaining thin and uniform coatings, fabricating multi-layer nanostructures on flexible substrate, producing seamless cylindrical imprint molds and roller mold lifetime, functional imprinting materials with fast curing, low viscosity, and shrinkage. To address these challenges, an infrastructure of large-area imprinting should firstly be established. **Figure 22** illustrated an infrastructure proposed by author. It mainly includes three aspects: the fundamental theory, the enabling technologies, and various industrial and potential applications. Practical demands form specific applications are main driving forces push large-area imprinting development and advances. The continuous and further research on fundamental theory and enabling technologies provide mass production capability for a variety of commercial applications.

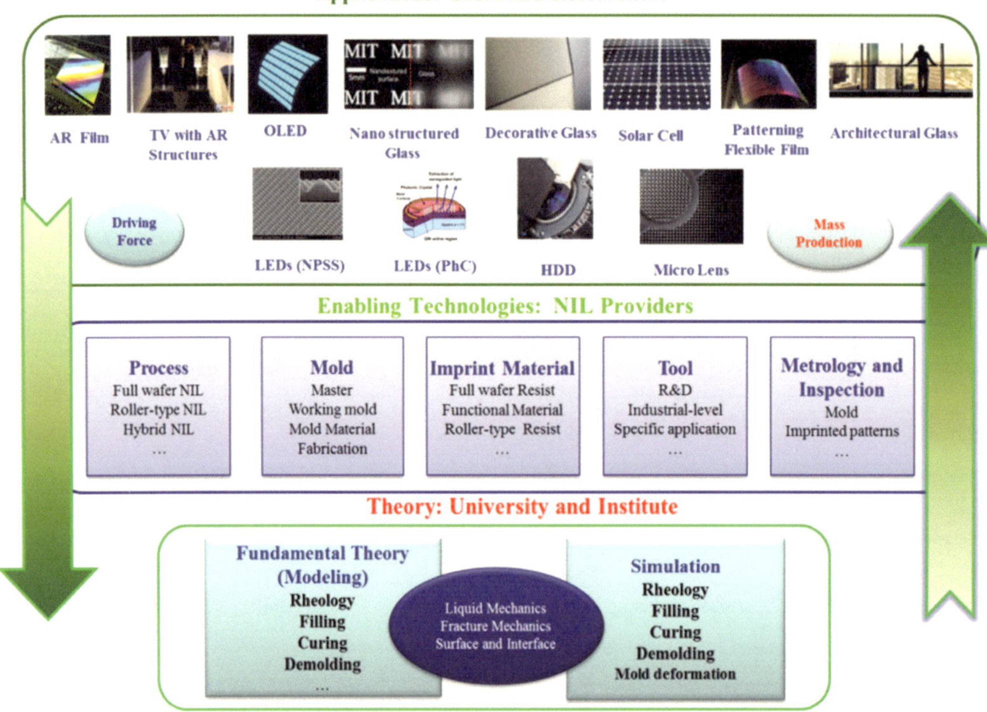

Figure 22. An infrastructure of large-area nanoimprinting.

Future trend for large-area imprinting process may be: film mold-based UV imprinting process, belt-type imprinting, R2R UV-NIL, fluoropolymer and PVA-based composite mold, large size seamless roller mold. Functional imprinting materials may be frequently used. Further enlarging imprinting area, improving resolution, and enhancing throughput and quality are needed.

5. Conclusions

NIL has over the two decade gone from being a new and exciting research topic, to be a technology used in the most advanced parts of various industries. Large-area NIL has been proven to be a cost-effective high volume nanopatterning method for large-area structure replication in the micrometer and nanometer scale, fabricating complex 3D micro/nanostructures, especially producing large-area patterns on the non-planar surfaces even curved substrates at low cost and with high throughput. In particular, it provides an ideal solution and a powerful tool for mass producing micro/nanostructures over large areas at low cost and high yield rate for the applications in HB-LEDs, anti-reflective coatings or films with moth-eye structures, flexible electronics, solar cells, architectural glass, optical elements, patterned media, micro-lens, OLED, wire grid polarizer, and functional polymer devices. That opens the way for many applications not previously conceptualized or economically feasible.

Full wafer NIL is a promising large-area wafer-scale nanoimprint method with nanoscale resolution. R2R NIL is attracting interest by both academic and industry around the world because of its inherent advantages of low cost, high throughput, and large-area patterning. Future advances in R2R nanoimprinting and its utilization in commercial applications strongly depends on the development and fabrication of new roller mold and resist materials that can maintain the fidelity of the replicated features and deliver an almost defect-free pattern transfer for long process cycles in a high volume manufacturing process. Unlike the full wafer NIL, during the R2R NIL process the mold material undergoes a severe test of its reliability, durability, robustness, mold recoverability, and the ability to maintain the fidelity of pattern transfer throughout the continuous process. Therefore the roller mold is one of the most important components that can determine the outcome of the entire process [35].

R2R NIL has been regarded as the closest process for the industrial application of NIL. In particular, the applications in anti-reflective coatings or films, flexible electronics, and wire grid polarizer have demonstrated significantly commercial prospect. Large-area NIL will become more and more important for many applications in large-area patterning, fabricating 3D micro/nanostructures and creating patterns on the non-planar or curved surface. There is a plenty of room to enhance the resolution, patterning area, mold lifetime, and yield for the promising patterning method.

Acknowledgements

Parts of this chapter are taken from the authors' former work (Soft UV Nanoimprint Lithography and Its Applications, and Nanoimprint Lithography, published by INTECH). This work was financially supported by National Science Foundation of China (Grant Nos. 51375250 and 51775288), and Qingdao Innovative Talents Project (13-CX-18).

Author details

Hongbo Lan[1,2]*

*Address all correspondence to: hblan99@126.com

1 Nanomanufacturing and Nano-Optoelectronics Lab, Qingdao University of Technology, Qingdao, China

2 Qindao Engineering Research Center for 3D Printing, Qingdao, China

References

[1] Verschuuren MA, Megens M, Ni Y, et al. Large area nanoimprint by substrate conformal imprint lithography (SCIL). Advanced Optical Technologies. 2017;**6**(3–4):243-264

[2] Bläsi B, Tucher N, Höhn O, et al. Large area patterning using interference and nanoimprint lithography. Proceedings of SPIE. 2016;**9888**:98880H

[3] Schoot JV, Schift H. Next-generation lithography—An outlook on EUV projection and nanoimprint. Advanced Optical Technologies. 2017;**6**(3–4):159-162

[4] Lan H, Ding Y, Liu H. Nanoimprint Lithography Principles, Processes and Materials. New York, USA: Nova Science Pub Inc.; 2011

[5] Dumond JJ, Low HY. Recent developments and design challenges in continuous roller micro- and nanoimprinting. Journal of Vacuum Science and Technology B. 2012;**30**(1):010801

[6] Ok JG, Shin YJ, Park HJ, et al. A step toward next-generation nanoimprint lithography: Extending productivity and applicability. Applied Physics A. 2015;**121**:343-356

[7] Viheriälä J, Niemi T, Kontio J, et al. Nanoimprint lithography—Next generation nanopatterning methods for nanophotonics fabrication. In: Kim K, editor. Recent Optical and Photonic Technologies. Rijeka: InTech; 2010. pp. P275-P298

[8] Yang Y, Mielczarek K, Aryal M, et al. Nanoimprinted polymer solar cell. ACS Nano. 2012;**6**(4):2877-2892

[9] Lee YC, Tu SH. Improving the light-emitting efficiency of GaN LEDs using nanoimprint lithography. In: Cui B, editor. Recent Advances in Nanofabrication Techniques and Applications. Rijeka: InTech; 2011. pp. 173-195

[10] Baek JH, Kim SM, Lee IH, et al. Control of characteristic performance by patterned structure in light-emitting diodes. Proceedings of SPIE. 2011;**7945**:79451B

[11] Lan H, Liu H. UV-nanoimprint lithography: Structure, materials and fabrication of flexible molds. Journal of Nanoscience and Nanotechnology. 2013;**13**(5):3145-3172

[12] Chou Y, Krauss P, Renstrom P. Imprint lithography with 25-nanometer resolution. Science. 1996;**272**(5258):85-87

[13] Schift H. Nanoimprint lithography: An old story in modern times? A review. Journal of Vacuum Science and Technology B. 2008;**26**(2):458-480

[14] Guo J. Nanoimprint lithography: Methods and material requirements. Advanced Materials. 2007;**19**(4):495-513

[15] Kooy M, Mohamed K, Pin L, et al. A review of roll-to-roll nanoimprint lithography. Nanoscale Research Letters. 2014;**9**:320

[16] Ahn SH, Guo LJ. Large-area roll-to-roll and roll-to-plate nanoimprint lithography: A step toward high-throughput application of continuous nanoimprinting. ASC. Nano. 2009;**3**(8):2304-2310

[17] Glinsner T, Plachetka U, Matthias T, et al. Soft UV-based nanoimprint lithography for large area imprinting applications. Proceedins of SPIE. 2007;**6517**:651718

[18] Verschuuren MA. Substrate conformal imprint lithography for nanophotonics [PhD thesis]. Utrecht University; 2011

[19] Ji R, Hornung M, Verschuuren MA, et al. UV enhanced substrate conformal imprint lithography (UV-SCIL) technique for photonic crystals patterning in LED manufacturing. Microelectronic Engineering. 2010;**87**:963-967

[20] Shinohara H, Fujiwara S, Tashiro T, et al. Formation of patterned sapphire substrate using UV imprint processes. Journal of Photopolymer Science and Technology. 2013;**26**(1):113-117

[21] Sato Y, Mizawa T, Mizukami Y, et al. Development of film mold for roll to roll nanoimprintg process and its application. Journal of Photopolymer Science and Technology. 2013;**26**(1):125-128

[22] Youn S, Ogiwara M, Goto H, et al. Prototype development of a roller imprint system and its application to large area polymer replication for a microstructured optical device. Journal of Materials Processing Technology. 2008;**202**:76-85

[23] Roller Press Scan® method. http://www.engineeringsystem.co.jp/automation/nanoimprint/en/ .2017/9

[24] Tan H, Gilbertson A, Chou S. Roller nanoimprint lithography. Journal of Vacuum Science and Technology B. 1998;**16**(6):3296-3298

[25] Ahn S, Cha J, Myung H, et al. Continuous ultraviolet roll nanoimprinting process for replicating large-scale nano- and micropatterns. Applied Physics Letters. 2006;**89**(21):213101

[26] Park S, Choi K, Kim G, et al. Nanoscale patterning with the double-layered soft cylindrical stamps by means of UV-nanoimprint lithography. Microelectronic Engineering. 2009;**86**(4–6): 604-607

[27] Nano-scale Pattern Productivity Enhanced 100 times by Sheet Nanoimprint Technology. Hitachi Research Laboratory. 2011

[28] Jiang W, Liu H, Ding Y, et al. Investigation of ink transfer in a roller-reversal imprint process. Journal of Micromechanics and Microengineering. 2009;**19**:015033

[29] Frenkel R, Kim B, Yao D. Extrusion roller imprinting with a Variotherrm belt mold. Machines. 2014;**2**(4):299-311

[30] Nagato K, Sugimoto S, Hamaguchi T, et al. Iterative roller imprint of multilayered nanostructures. Microelectronic Engineering. 2010;**87**:1543-1545

[31] EVG®750R2R Automated Hot Embossing System. http://www.evgroup.com. 2017

[32] Ahn BS, Guo LJ. High-speed roll-to-roll nanoimprint lithography on flexible plastic substrates. Advanced Materials. 2008;**20**:2044-2049

[33] Jeans A, Almanza-Workman M, Cobene R, et al. Advances in roll-to-roll imprint lithography for display applications. Proceedings of SPIE. 2010;**7637**:763719

[34] Kim H, Almanza-Workman M, Garcia B, et al. Roll-to-roll manufacturing of electronics on flexible substrates using self-aligned imprint lithography (SAIL). Journal of the SID. 2009;**17**:963-970

[35] John J, Tang Y, Rothstein J, et al. Large-area, continuous roll-to-roll nanoimprinting with PFPE composite molds. Nanotechnology. 2013;**24**:505307

[36] Inanami R, Ojima T, Matsuki K, et al. Sub-100 nm pattern formation by roll-to-roll nanoimprint. Proceeding of SPIE. 2012;**8323**:83231J

[37] Ok JG, Ahn SH, Kwak MK, et al. Continuous and high-throughput nanopatterning methodologies based on mechanical deformation. Journal of Materials Chemistry C. 2013;**1**:7681-7691

[38] Burghoorn M, Roosen-Melsen D, Riet J, et al. Single layer broadband anti-reflective coatings for plastic substrates produced by full wafer and roll-to-roll step-and-flash nano-imprint lithography. Materials. 2013;**6**:3710-3726

[39] Zhong ZW, Shan XC. Microstructure formation via roll-to-roll UV embossing using a flexible mould made from a laminated polymer-copper film. Journal of Micromechanics and Microengineering. 2012;**22**:085010

[40] Byeon KJ, Hong E, Park H, et al. Full wafer scale nanoimprint lithography for GaN-based light-emitting diodes. Thin Solid Films. 2011;**519**:2241-2246

[41] Lim H, Choi K, Kim G, et al. Roller nanoimprint lithography for flexible electronic devices of a sub-micron scale. Microelectronic Engineering. 2011;**88**:2017-2020

[42] Yang Y, Mielczarek K, Zakhidov A, et al. Efficient low bandgap polymer solar cell with ordered heterojunction defined by nanoimprint lithography. ACS Applied Materials & Interfaces. 2014;**6**:19282-19287

[43] Lee KT, Jang JY, Zhang J, et al. Highly efficient colored perovskite solar cells integrated with ultrathin subwavelength plasmonic nanoresonators. Scientific Reports. 2017;**7**:10640

Chapter 4

Micro/Nano Patterning on Polymers Using Soft Lithography Technique

Sujatha Lakshminarayanan

Additional information is available at the end of the chapter

http://dx.doi.org/10.5772/intechopen.72885

Abstract

Microfabrication is essential in the field of science and technology. The development and innovations in this field are already prominent in the society through microelectronics and optoelectronics. The lithography or transfer of pattern to the substrate/surface of a layer is an important process step in microfabrication and is usually carried out with photolithography. Though photolithography is a well-established technique, it suffers from drawbacks such as limited feature size due to optical diffraction, requirement of high-energy radiation for small features, and high-cost involvement for sophisticated instruments. Also, it cannot be applied to nonplanar surfaces. Soft lithography is complement to photolithography which overcomes the above-mentioned drawbacks. Soft lithography is a simple and inexpensive method, and also, it suits to wide range of materials and very large surface areas. High-quality micropatterns or nanopatterns can be made using a patterned elastomeric stamp. This article briefly describes the various soft lithography techniques to obtain high-resolution structures for nanofabrication.

Keywords: soft lithography, polydimethylsiloxane, SU8, UV lithography, photoresist

1. Introduction

Microfabrication is essential in the field of science and technology. The development and innovations in this field are already prominent in the society through microelectronics and optoelectronics. The lithography or transfer of pattern to the substrate/surface of a layer required in microfabrication is usually carried out with photolithography. Photolithography is the basic technology used for making all microelectronic systems. Photolithography is limited to materials for various layers or substrates due to the etching chemistry. Also, photolithography is limited to geometries it can produce. At the same time, photolithography technique is expensive and can only pattern a small area at any given time. Also, the feature

size of the pattern is limited by diffraction of light. As photolithography is confined to extremely flat silicon substrates, one cannot fabricate electronic circuits on a plastic sheet or curved surface. Soft lithography is an alternate technology which provides a good control over an infinite range of structures and chemistries. The dimension can be defined from nanoscale to mesoscale, and it finds applications in developing devices in different fields from consumer product to life sciences and industrial processes.

Conventional photolithography has its own disadvantages such as the following:

i. The feature size is limited by optical diffraction

ii. Complex facilities and technologies for high-energy radiation needed for small feature sizes (EUV, E-beam, or X-ray lithography techniques)

iii. The system is very expensive

iv. Not suitable for nonplanar surfaces

v. No control over the chemistry of patterned surfaces

Due to the above difficulties in photolithography, nonphotolithography technique called soft lithography came into picture, which overcomes the above problems. It does not have the diffraction limits and provides access to three-dimensional structures, chemically inert, inexpensive, and simple processes suitable for molecular scientists. A high quality of patterns with lateral dimensions of about 30 nm–500 μm can be obtained using this technique. Soft lithography is a complement to conventional lithography system and has numerous advantages such as possibility to pattern UV sensitive materials without degrading the performance, to pattern nonplanar surfaces, to pattern large surfaces, to control the chemistry during patterning, ideally does not have any diffraction limits, short time between idea and prototype, clean room free operation, etc.

This chapter explores the fabrication of micromolds with polydimethylsiloxane (PDMS). PDMS is a Si-based organic polymer that has found wide applications in MEMS for soft lithography. PDMS has several desirable properties, which are:

i. Visco-elasticity: PDMS is flexible

ii. Biocompatibility

iii. High chemical inertness

iv. Optical transparency

v. Adhesion to metals: Applications as inert substrate material

Initially, the fabrication of polymer molds has been investigated to explore the nonconventional lithography technology called "soft lithography." As this technique requires the cycle time of less than 24 hours from design to implementation, many researchers are using this technique for rapid prototype development. The use of polymers for microfluidics is the major research area in the field of Medical Diagonostics and Tissue Engineering. Soft lithography using PDMS molds had been first developed by Whitesides in 1998 at Harvard University [1, 2].

The disadvantages of conventional optical lithography and the advantages of soft lithography had been well demonstrated by his team. He also established the various soft lithography techniques to get the feature size in the range of 30 nm to 500 μm. Many researchers reported on the various techniques of soft lithography in detail [3–8]. Research has been carried out to develop various microdevices using this technique in the field of optics and biosensors. Huang et al. reported about the replication of polymeric microring optical resonators with soft lithography and found an excellent agreement in the optical properties between molded replicated devices and master devices [9]. Chang-Yen reported on PDMS waveguide system using soft lithography [10]. The attenuation and temporal stability were excellent at low cost, low toxicity, and biocompatibility with yield of 96%. Golden et al. reported on array biosensor with PDMS mold which can be used for the simultaneous detection of multiple targets in multiple samples within 15–30 minutes [11]. Liu et al. have reported the fabrication of microchannels, in which with the aid of Si-reinforced PDMS molds, Au dots and wires have been successfully generated on the sidewalls and bottom surfaces of microchannels through hot-embossing processes [12]. Tarbague et al. discussed about the development of new PDMS microfluidic chip molding for Long Wave Biosensor to realize a detection cell of bio-organisms in liquid media [13].

2. Various methods of soft lithography

There are various methods of soft lithography to get precise micropatterns or nanopatterns on the planar or nonplanar surfaces. The following soft lithography process methods have been well established for various commercial devices: (i) replica mold (REM) technique (ii) Microcontact printing (μCP) technique (iii) Micromolding in capillaries (MIMIC) (iv) Solvent-assisted micromolding (SAMIM) (v) Microtransfer molding (μTM) and (vi) Hot embossing technique.

2.1. Replica mold (REM) technique

Replica molding is a very old, simple, and reliable method, in which the micro- or nanopatterns on the surface of the prime (master) mold is duplicated on the polymer material [14]. The minimum feature size of less than 100 nm can be accurately replicated using this technique. It has been successfully used for mass production of devices such as compact disks (CDs) [15, 16], diffraction gratings [17], holograms [18], and microtools [19]. The prime or master mold is generally fabricated on a rigid material (silicon, nickel, glass, or SU8 photoresist) using a standard photolithography and micromachining techniques. The elastomeric stamp or replica mold is fabricated by cast-molding technique. A prepolymer of the elastomer is poured over the master, cured, and then peeled off. Poly(dimethylsiloxane) (PDMS) is a widely used elastomer all over the globe compared to other elastomers such as polyurethanes, polyimides, and Novolac resins. PDMS is a unique combination of an inorganic siloxane and organic methyl groups. Since the glass transition temperature of PDMS is very low, it is available in the form of fluid at room temperature. This can be readily converted into solid elastomers by cross-linking. Nowadays, PDMS is readily available in the market as a two-part material containing prepolymer and curing agent.

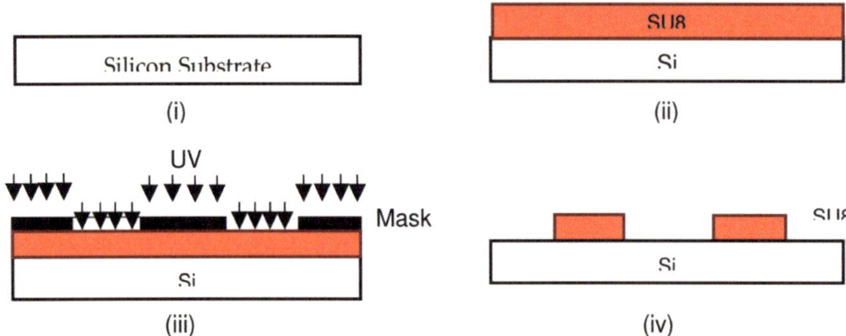

Figure 1. Process flow of master mold fabrication using UV lithography. (i) Silicon substrate (ii) SU8 is spin coated on silicon wafer for required thickness and (iii) exposure of UV light and (iv) pattern obtained after developing.

Figure 1 shows the process flow for the fabrication of master mold. The master is made up of SU8–2075 negative photoresist (Microchem, USA), which helps us to get high-aspect ratio structures precisely. SU8–2075 is spin coated on flat silicon substrate for the required thickness. For example, to get the photoresist thickness of 100 μm, spinning was carried out with the spin speed of 2230 rpm for spin time of 30 sec with an acceleration of 500 rpm for 5 sec. The sample was soft baked for 5 minutes at 65°C then for 12 minutes at 95°C. Now, the mask plate (written by laser writing on Chrome coated glass plate) was kept above the sample, and UV light (365 nm wavelength) was exposed for 15 seconds using Karl Suss Mask Aligner. Then, the sample was post baked for 5 minutes at 65°C then for 10 minutes at 95°C. Then, it was developed for 7 minutes using SU8 developer solution. Finally, the sample was hard baked for 1 hour at 95°C, and the patterns on SU8 photoresist were observed under the microscope. The master made on SU8 can be used repeatedly for more than 50 times to make polymer replicas.

Figure 2 shows the process flow for the fabrication of PDMS stamps using the master. The elastomeric stamp or mold is prepared by cast molding. PDMS elastomer Sylgard 184 was

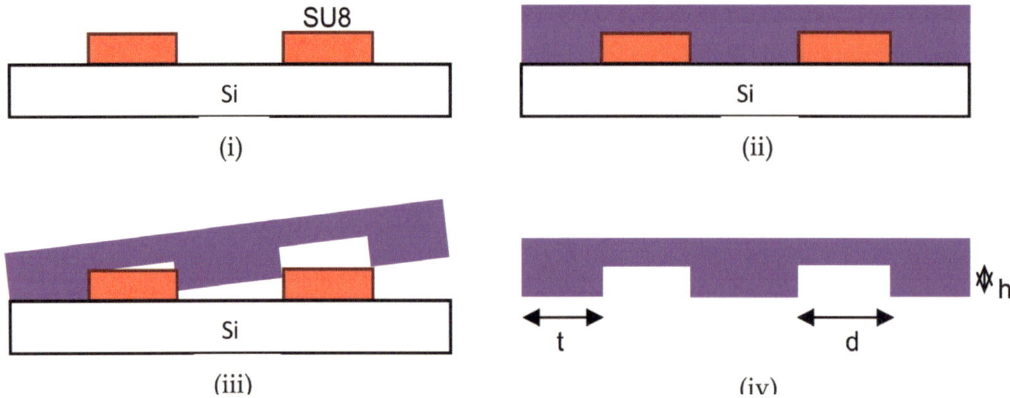

Figure 2. Process flow of PDMS stamps fabrication (i) SU8 master mold on silicon substrate (ii) PDMS prepolymer poured on SU8 (iii) peeling off PDMS replica after curing (iv) fabricated PDMS stamp.

obtained from Dow Corning. It comes as a two-part material containing the silicone base and a curing agent. The base and the curing agent were taken in the ratio of 10:1, and vigorous agitation by manual stirring is required for mixing. During this process, air bubbles may form in this mixture. To remove the air bubbles, the mixture was kept in vacuum desiccators for 1 hour. Then, it was poured on the SU8 master mold structure and cured at 70°C for 3 hours. After cooling down to room temperature, the PDMS layer was peeled off from SU8 pattern. The resulting PDMS micromold pattern is a reverse pattern of SU8 structure. **Figure 3** shows the SEM images of three different master molds made up of SU8–2075, and **Figure 4** shows their corresponding PDMS replica patterns.

Replica molding is extensively used to fabricate microfluidic devices as a rapid prototype [20–25]. The microchannels are formed on the PDMS mold with SU8 as prime mold. Then, this PDMS structure is placed on a glass cover plate and bonded together by oxygen plasma treatment of both surfaces. The PDMS is punched to form inlet and outlet ports, and tygon tubes with luer connectors are inserted on the ports. The photograph of the microfluidic device fabricated using the above method is shown in **Figure 5**.

Replica molding technique has been adapted for the fabrication of topologically complex, optically functional surfaces that would be difficult to fabricate with other techniques. Experimenting with various feature size of photo lithography and test the replica molding for its best replication is also necessary. PDMS shrinks by approximately 1% upon curing. Also,

Figure 3. SEM images of three different SU8 master mold patterns (i), (ii), and (iii).

Figure 4. SEM images of three different PDMS mold patterns: (i)–(iii) corresponding to SU8 master mold patterns shown in **Figure 3** (i)–(iii).

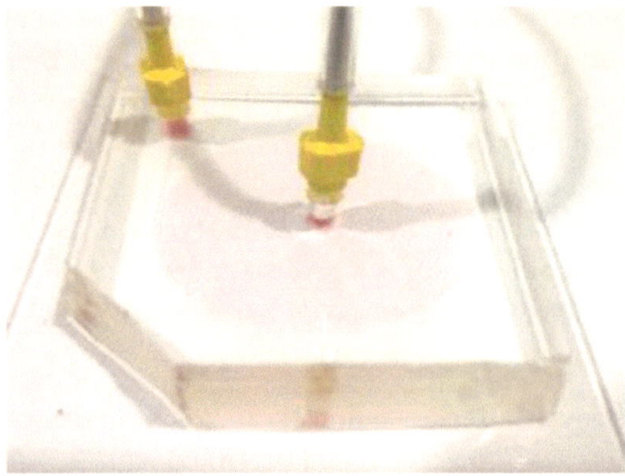

Figure 5. Photograph of the spiral microfluidic device fabricated using replica molding on PDMS.

the cured PDMS gets swelled while treating with organic solvents such as toluene and hexane. Due to low Young's Modulus and thermal expansion of PDMS, there are variations in the dimensions especially while working with multilayered structures. Also, while fabricating PDMS with microstructures, if the aspect ratio (*h*/*l*) is too high or too low, the elastomeric character of PDMS will cause the defects in the microstructures as shown in **Figure 6**. During fabrication, stress is induced on PDMS due to gravity, adhesion, or capillary forces and generates defects in the pattern that is formed. Main technical problems are given below [1]:

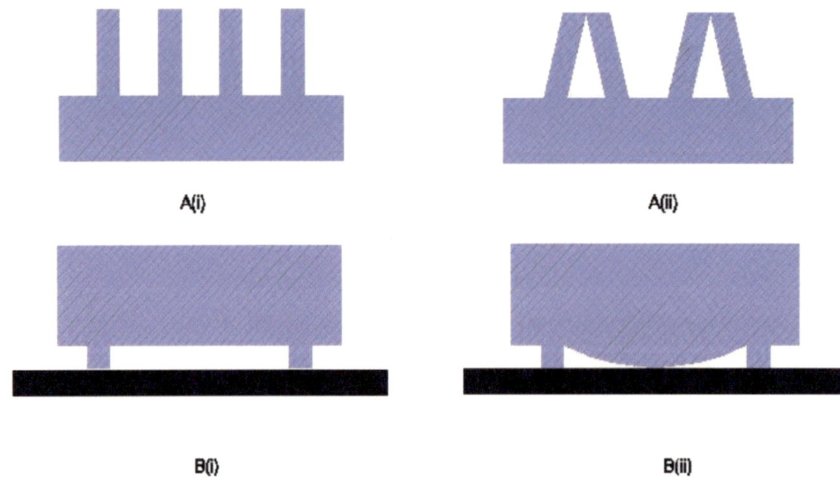

Figure 6. Defects in patterns due to high or low aspect ratio structures: A(i) expected structure with high aspect ratio (ii) resultant collapsed structure; B(i) expected low aspect ratio structure with (ii) resultant sagged structure.

i. If the aspect ratio is too large, the PDMS microstructures fall under their own weight as shown in **Figure 6**. **Figure 6** A(i) shows the expected structure, and A(ii) shows the resultant structure due to high aspect ratio. Aspect ratios between 0.2 and 2 are the best to get defect-free stamps.

ii. If the aspect ratio is too small, the relief structures are not able to withstand the compress forces and adhesion between the stamp and the substrate. This will lead to sagging as shown in **Figure 6** B(i) and (ii). Nonfunctional posts or rigid supports have to be included in the design to avoid sagging.

2.2. Microcontact print (μCP) technique

Microcontact printing (μCP) is a simple and efficient method for precise microscale pattern transfer for biotechnology applications. This technique enables tailoring the properties of surface at molecular level using self-assembled monolayers (SAMs). SAMs can be easily formed by immersion of the stamp in the solution containing a ligand ($Y(CH_2)_nX$) reactive toward the surface or by exposure of the stamp to the vapor of a reactive species [2]. The thickness of a SAM can be controlled by change in the number (n) of methylene groups in the alkyl chain. Changing the head group X can modify the surface of the monolayer. The binding of the anchoring group Y is selective to substrate material. Well-established methods of SAMs of alkanethiolates on Au and Ag and alkylsiloxanes on hydroxyl-terminated surfaces such as Si/SiO_2, Al/Al_2O_3, glass, mica, and plasma-treated polymers are reported.

Figure 7 shows the process steps for microcontact printing. A thin metal film such as gold (Au), silver (Ag), copper (Cu), palladium (Pd), or platinum (Pt) is deposited on the substrate by physical vapor deposition (thermal evaporation or e-beam evaporation). PDMS stamp (as shown in **Figure 7(a)**) is wetted with a hexadecanethiol in ethanol (as shown in **Figure 7(b)**) and is brought into contact with the surface of Au for 10:20 seconds. The hexadecanethiol transfers from the stamp to the gold upon contact resulting in a hexadecanethiolate and generates patterns of SAMs on the surface of gold as shown in **Figure 7(c)**.

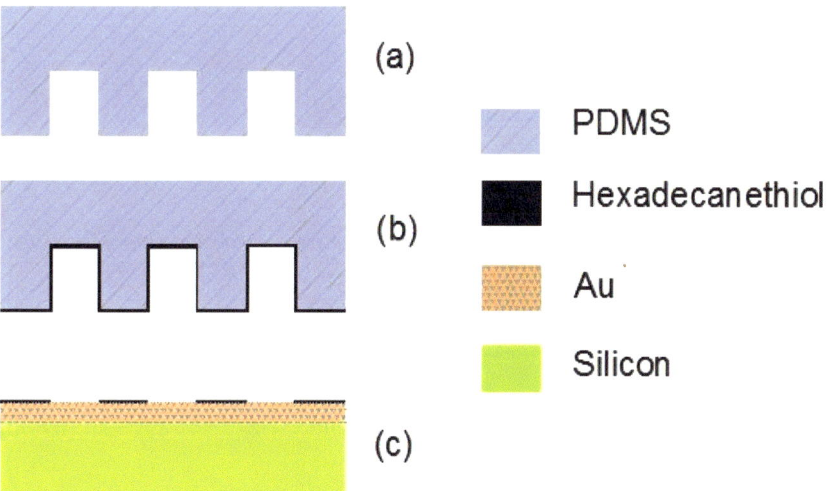

Figure 7. Process steps for microcontact printing (a) PDMS stamp (b) PDMS stamp immersed in hexadecanethiol solution and (c) SAM pattern transferred to the substrate with gold coating by bringing the stamp in contact upon the substrate.

2.3. Micromolding in capillaries (MIMIC)

Micromolding in capillaries (MIMIC) is another technique of generating microstructures using PDMS stamps. In this method, top portion of microchannel structures is made on the PDMS stamp. Then, this stamp is placed on a substrate with the channel structure facing down on the substrate. This forms capillaries on the substrate. When a polymer material such as polyurethane or epoxy is placed at one end of the channel, it started flowing into the channels due to capillary action. After some time, the capillaries are filled by the above material. Finally, this polymer material is cured by UV or heat or using a curing agent. Once the polymer is cured, the PDMS stamp can be removed. The fill rate of the polymer inside the capillary depends on viscosity of the fluid, radius of the channel, pressure difference at the ends of the capillaries, surface tension, contact angle, and the length of the capillary. **Figure 8** shows the process flow for micromolding in capillaries (MIMIC).

2.4. Solvent-assisted micromolding (SAMIM)

In solvent-assisted micromolding (SAMIM) technique, a PDMS mold with microchannels is placed on a substrate. As PDMS is hydrophobic in nature, it is treated with oxygen plasma to improve its wettability. A good solvent that can dissolve a polymeric material without affecting PDMS mold is allowed to fill the microchannels. The solvent gets evaporated and the resulting fluid or gel which solidifies and made available in the molded structures defined by the PDMS mold. The solvents such as methanol, ethanol, Iso-propane alcohol, toluene, and acetone having high surface tension are generally being used. The problem with SAMIM is the formation of a very thin film at the bottom of the structure. This film has to be removed by RIE or by a suitable etchant before use for the preferred application. **Figure 9** shows the step-by-step process flow for the SAMIM technique of patterning.

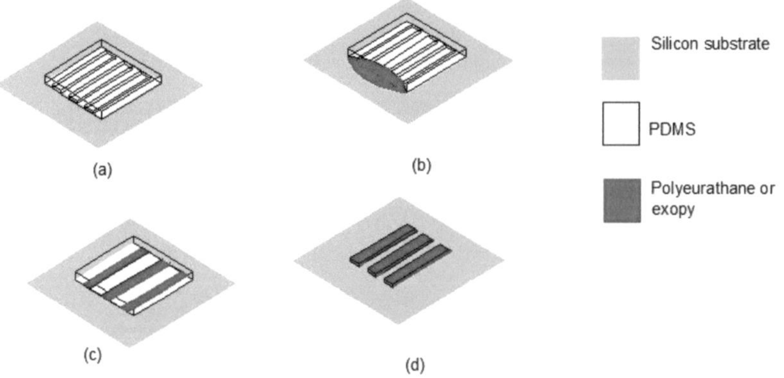

Figure 8. Process flow for micromolding in capillaries (a) PDMS mold with microchannels on silicon substrate (b) Polyeurathane or epoxy in the form of fluid at the end of the microchannels (c) fluid flows inside the microchannels by capillary force and (d) cured epoxy and removal of the PDMS mold.

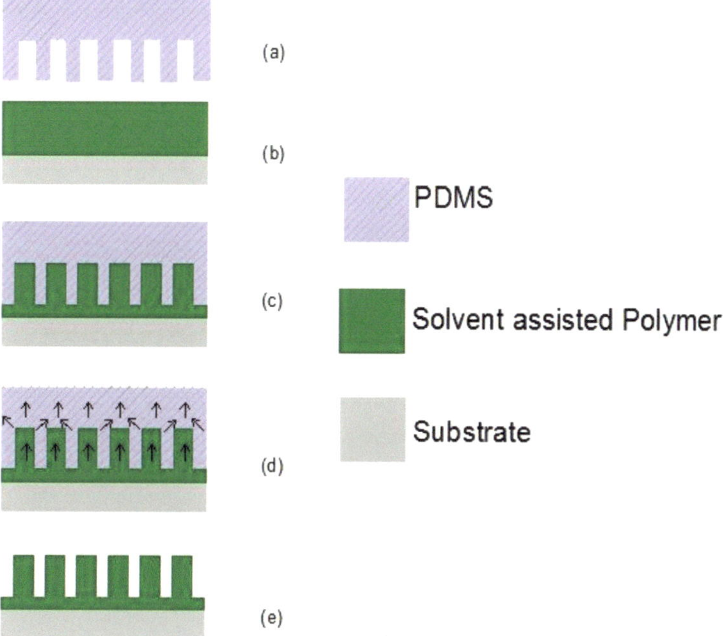

Figure 9. Process steps for SAMIM (a) PDMS mold (b) solvent assisted polymer on a substrate (c) polymer mold placed on the solvent (d) solvent evaporates and (e) microstructures of solidified polymer after solvent evaporation.

2.5. Microtransfer molding (μTM)

Microtransfer molding (μTM) is a simple soft lithography technique to form patterned microstructures of polymers such as organic polymers or polyeurathane on a large area. The polymer may also be doped with florescent material like rhodomine. In this method, the patterned surface of a PDMS mold is filled with a prepolymer of required polymer material as shown in **Figure 10(a)**. The excess fluid is removed by another flat PDMS block to get a flat surface as shown in **Figure 10(b)**. This combined block is turned upside down and placed on a substrate as shown in **Figure 10(c)**. Now, the prepolymer is cured by UV light or by heating. Once the prepolymer is solidified, the PDMS mold is peeled out leaving the patterned polymer structure on the surface of the substrate as shown in **Figure 10(d)**. In this method, a very thin layer (100 nm) of polymer is formed in-between the raised structure. This can be etched by reactive ion etching (RIE). This method is established for the fabrication of optical waveguides, couplers, and interferometers.

2.6. Hot embossing

Hot embossing also referred as soft embossing denotes stamping of micropatterns onto a polymer softened by raising the temperature just above its glass transition temperature. The stamp used to define the pattern may be made up of a hard material such as Silicon or metal

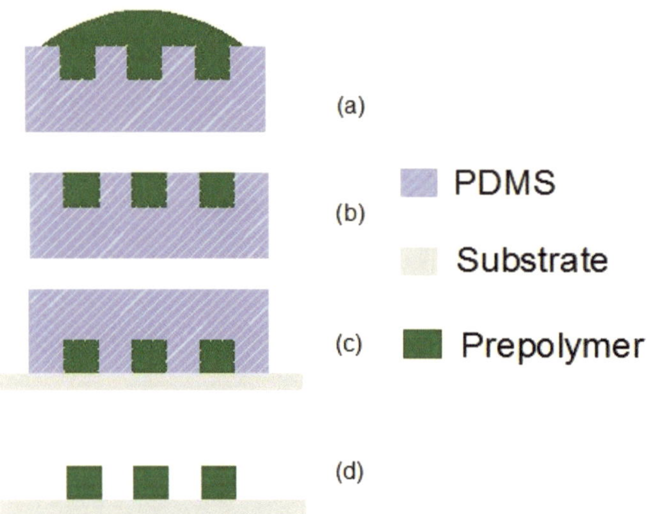

Figure 10. Process flow for the fabrication of polymer microstructures using microtransfer molding technique (a) PDMS mold filled with prepolymer (b) excess fluid is removed and the surface is flattened (c) the mold with prepolymer is kept on the substrate with upside down (d) Prepolymer is cured and the PDMS mold is peeled off.

using any of the micromachining technique like bulk silicon etching or LIGA. The preferred polymer materials suitable for making the replicas using this technique may be polystyrene (PS) or poly methyl methacrylate (PMMA) or polycarbonate (PC). In this method, dimensions of micropatterns less than 1 μm are highly possible.

The process involves three steps namely heating, embossing, and demolding as shown in **Figure 11**. During the first step of heating, the metal or silicon mold and the polymer substrate are aligned and placed in between two stainless steel supporting plates. This setup is brought into a heating chamber, and the sample is kept at the specified embossing temperature for a soaking time of 30 minutes. During the second step of embossing, the load is applied gradually by a pneumatic (servomotor controlled) press and hold at a specific load for few minutes. This is followed by gradual unloading. The third step is demolding, in which the temperature is gradually reduced. After it is cooled down, the mold and the polymer replica are taken out of

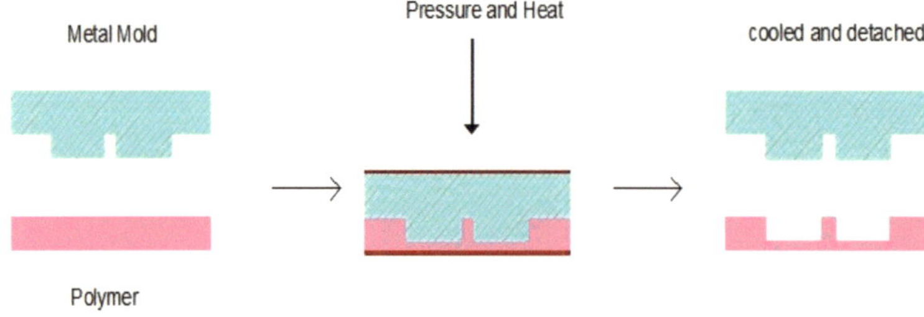

Figure 11. Process flow of hot embossing technique: (a) Step 1: prepare the metal mold and keep over the required polymer sheet; (b) Step 2: apply pressure and heat for hot embossing; (c) Step 3: cool and detach the polymer replica.

the chamber and separated out. For better precision, a vacuum chamber is preferred which maintains the temperature and helps for perfect embossing of microstructures.

3. Challenges

Though soft lithography seems to be simple and promising low-cost technique to achieve nanostructures, there are many challenges in bringing this technique to market. The major problem is the distortion of elastomeric materials, which limits high-resolution registration in soft lithography. This problem can be reduced by using thick samples and rigid supporting structures. The micropatterns or nanopatterns in the stamp or mold may distort due to pairing, sagging, swelling, and shrinking of elastomer. The process may also introduce defects due to dust particles, poor adhesion to substrate, or poor release from the stamp and bubbles in the prepolymer. The presence of thin film of polymer in soft lithography is generally removed by reactive ion etching. But, this may damage small features. Soft lithography is still in its early stage of development, and researchers and manufacturers of microdevices are working in establishing this technology toward reliability, reproducibility, and stability of the micro/nanostructures.

Acknowledgements

The author acknowledges Defense Research and Development Organization (ER&IPR), Government of India (No. ERIP/ER/1204662/M/01/1498) for the financial assistance to carry out research works on fabrication of microfluidic devices using soft lithography technique.

Author details

Sujatha Lakshminarayanan

Address all correspondence to: sujatha.l@rajalakshmi.edu.in

Centre of Excellence in MEMS and Microfluidics, Rajalakshmi Engineering College, Chennai, India

References

[1] Xia Y, Whitesides GM. Reviews: Soft lithography. Angewandte Chemie, International Edition. 1998;**37**:550-575 DOI: 1433-7851/98/3705-0551

[2] Xia Y, George M. Whitesides, soft lithography. Annual Review of Material Science. 1998;**28**:153-184 DOI: 0084-6600/98/0801-0153

[3] Christopher RM, Aksay IA. Microchannel molding: A soft lithography-inspired approach to micro-meter scale patterning. Journal of Materials Research. 2005;**20**(8):1995-2003. DOI: 10.1557/JMR.2005.0251

[4] Cooper MDonald J, Duffy DC, Anderson JR, Chiu DT, Wu H, Schueller OJA, George M. Whitesides, review: Fabrication of microfluidic systems in poly(dimethylsiloxane). Electrophoresis. 2000;**21**:27-40. DOI: 10.1002/(SICI)1522-2683(20000101)21:1<27::AID-ELPS27>3.0.CO;2-C

[5] Rogers JA, Nuzzo RG. Recent progress in soft lithography. Materials Today. 2005:50-56. DOI: 10.1016/S1369-7021(05)00702-9

[6] Qin D, Xia Y, Whitesides GM. Soft lithography for micro and nanoscale patterning. Nature Protocols. 2010;**5**(3):491-502. DOI: 10.1038/nprot.2009.234

[7] Krogh M. My Little Guide to Soft Lithography. Sweden: Linkoping University; www.ifm.liu.se/applphys/biorgel/education/.../Soft_Lithography_for_Dummies.pdf

[8] Kang S-W. Application of soft lithography for Nano functional devices. Chap 20. In: Wang M, editor. Lithography. (Ed.), ISBN: 978-953-307-064-3. InTech; 2010. DOI: 10.5772/8186

[9] Huang Y, Paloczi GT, Scheuer J, Yariv A. Soft lithography replication of replication of polymeric microring optical resonators. Optics Express. 2003;**11**(20):2452-2458. DOI: 10.1364/OE.11.002452

[10] Chang-Yen DA. A monolithic PDMS waveguide system fabricated using soft-lithography techniques. Journal of Lightwave Technology. 2005;**23**:2088-2093. DOI: 10.1109/JLT.2005.849932

[11] Golden J, Shriver-Lake L, Sapsford K, Ligler F. A "do-it-yourself" array biosensor. Methods. 2005;**37**:65-72. DOI: 10.1016/j.ymeth.2005.05.010

[12] Liu X, Luo C. Fabrication of au sidewall micropatterns using Si-reinforced PDMS molds. Sensors and Actuators A. 2009;**152**:96-103. DOI: 10.1016/j.sna.2009.03.016

[13] Tarbague H, Lachaud JL, Destor S, Vellutini L, Pillot JP, Bennetau B, Pascal E, Moynel D, Mossalay D, Rebiere D, Dejous C. PDMS microfluidic chip molding for love wave biosensor. Journal Integrated Circuits and Systems. 2010;**5**(2):125-133

[14] Hemling M, Crooks JA, Oliver PM, Brenner K, Gilbertson J, Lisensky GC, Weibel DB. Microfluidics for high school chemistry students. Journal of Chemical Education. 2014;**91**:112-115. DOI: 10.1021/ed4003018

[15] Sarkar S, Biswas PK, Jana S. Nano silver coated patterned silica thin film by sol-gel based soft lithography technique. Journal of Sol-Gel Science and Technology. 2012;**61**:577-584 doi.org/10.1007/s10971-011-2663-9

[16] Terris BD, Mamin HJ, Best ME, Logan JA, Rugar D. Nanoscale replication for scanning probe data storage. Applied Physics Letters. 1996,1997:**69**:4262-4264. https://doi.org/10.1063/1.116965

[17] Ramos BL, Choquette SJ. Embossable grating couplers for planar waveguide optical sensors. Analytical Chemistry. 1996;**68**:1245-1249. DOI: 10.1021/ac950579x

[18] Moon JH, Small A, Yi G-R, Lee S-K, Chang W-S, Pine DJ, Yang S-M. Patterned polymer photonic crystals using soft lithography and holographic lithography. Synthetic Metals. 2005;**148**:99-102. DOI: 10.1016 / j.synthmet.2004.09.019

[19] Cavallini M, Gentili D, Greco P, Valle F, Biscarini F. Preparation of tools for lithographically controlled wetting and soft lithography. Protocol Exchange. 2012. DOI: 10.1038/protex.2012.029

[20] Yang, Jiang. Elasomer application in microsystem and microfluidics. Chap 8, Advanced Elastomers – Technology, Properties and Applications; 2012. Intech from https://cdn.intechopen.com/pdfs-wm/38883.pdf

[21] Khademhosseini A, Suh KY, Jon S, Eng G, Yeh J, Chen G-J, Langer R. A soft lithographic approach to fabricate patterned microfluidic channels. Analytical Chemistry. 2004;**76**: 3675-3681. DOI: 10.1021/ac035415s

[22] Kim P, Kwon KW, Park MC, Lee SH, Kim SM, Suh KY. Soft lithography for microfluidics: A review. BioChip Journal. 2009;**2**(1):1-5. DOI: 10.1.1.458.853

[23] Abdelgawad M, Wu C, Chien W-Y, Geddie WR, Jewett MAS, Sun Y. A fast and simple method to fabricate circular microchannels in polydimethysiloxane (PDMS). Lab on a Chip. 2011;**11**:545-551. DOI: 10.1039/c0lc00093k

[24] Karthikeyan K, Sujatha L. Fabrication of microfluidic devices using softlithography technique. Proc. of SENSOR 2016 Conference 2016 on 29–30 February 2016 at RCI, Hyderabad

[25] Karthikeyan K, Sujatha L, Design and fabrication of microfluidic device for mercury ions detection in water, International Conference on NextGen Electronic Technologies: Silicon to Software, 23–25 March 2017 IEEExplore. http://ieeexplore.ieee.org/document/8067934/

Chapter 5

EUV/Soft X-Ray Interference Lithography

Shumin Yang and Yanqing Wu

Additional information is available at the end of the chapter

http://dx.doi.org/10.5772/intechopen.74564

Abstract

Based on the coherent radiation from an undulator source, extreme UV interference lithography (EUV-IL) technology is considered as the leading candidate for future nodes of high-volume semiconductor manufacturing. The throughput of this technique is much higher than that of traditional lithography methods such as e-beam lithography (EBL) and laser interference lithography (LIL). Different types of interference schemes based on reflection mirrors and transmission diffraction masks have been described in this chapter. Achromatic Talbot lithography (ATL) and the soft X-ray interference lithography (SXIL) with different photon energies have also been developed to produce highly dense, high-resolution periodic nanostructures. Two scan-exposure techniques, one is the method employing the broadband Talbot effect and the other based on the multi-grating EUV-IL with an order sorting aperture (OSA), have been used to obtain periodic nanostructures over large areas. Applications of EUV-IL on EUV-resist testing and nano-science have been illustrated.

Keywords: EUV, interference lithography, achromatic Talbot lithography, soft X-ray interference lithography, periodic nanostructures

1. Introduction

1.1. EUV interference lithography

Based on the coherent radiation from an undulator source, extreme ultraviolet interference lithography (EUV-IL) has been proven to be a powerful tool for high-resolution periodic nanostructure fabrication. The throughput of this technique is much higher than that of traditional lithography methods such as e-beam lithography (EBL) and laser interference lithography (LIL). Based on the interference of two or more coherent beams, interference lithography (IL) is usually used as a simple

method for large-area periodic nanostructure fabrication. Through a combination of advantages of IL and the short wavelength of EUV, EUV-IL technology has been proved as a powerful tool for high-resolution nanostructure fabrication over large areas. So far, several EUV-IL beamlines have been built in different synchrotron facilities, such as the XILII beamline in swiss light source (SLS) [1], the EUV-IL beamline in shanghai synchrotron radiation facility (SSRF) [2], the EUV-IL beamline in New SUBARU synchrotron radiation facility [3], the EUV-IL beamline at the University of Wisconsin-Madison [4] and the EUV-IL beamline in the national synchrotron radiation research center (NSRRC), Taiwan [5]. Up to now, the highest-resolution line structures with 6-nm half pitch (HP) have been afforded in PSI XILII [1]. During the last decade, a number of interference schemes were investigated including Lloyd's mirror, two-grating and multi-grating schemes.

1.2. Different types of EUV-IL

1.2.1. Interference with reflection optics

Figure 1a illustrates the general scheme of a Lloyd's mirror interferometer. The single-plane mirror is the key component in the interference scheme. A part of the incident beam is reflected by this mirror which interferes with the unreflected part of the beam to form interference fringes. Line structures with periods as small as 38 nm have been fabricated by this method at the Synchrotron Radiation Center (SRC), as shown in **Figure 1b** [6, 7]. Mirrors coated by the resonance multilayer are often used to split and reflect the soft X-rays. The extremely high requirements on mirror surface quality and the limited coherence of the soft X-ray sources complicate the implementation of such interferometers.

1.2.2. Interference with diffraction optics

Figure 2 illustrates the general scheme of the IL method with transmission diffraction optics. Under normal illumination from a spatially coherent EUV source, diffracted beams through the transmission mask interfere together at a certain distance from the mask. Line-space structures and 2D periodic structures will be fabricated by the two- and four-beam transmission

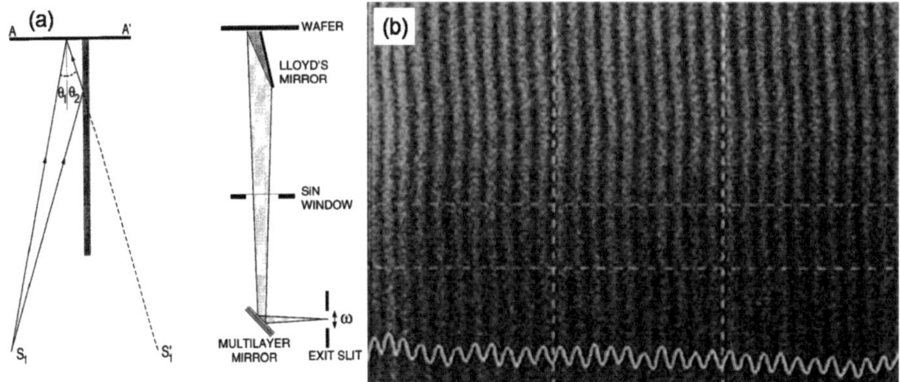

Figure 1. (a) The general scheme of a Lloyd's mirror interferometer. (b) Line structures with 38-nm period fabricated by SRC using the Lloyd's mirror interferometer (reproduced from [6], with the permission of AIP publishing).

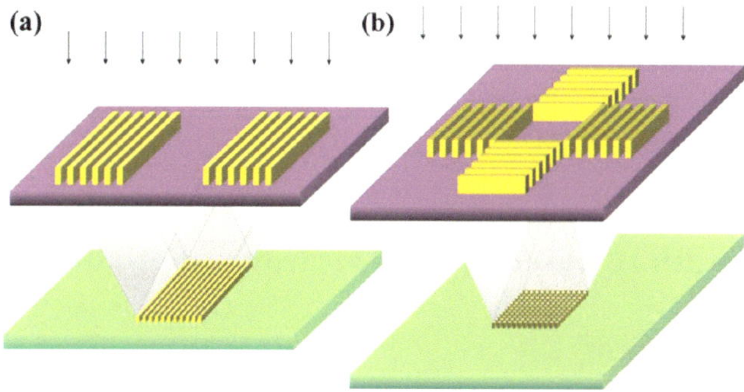

Figure 2. The general scheme of the two-beam (a) and four-beam (b) transmission-diffraction IL method (reproduced with permission from [8] ©(2009) COPYRIGHT Society of Photo-Optical Instrumentation Engineers (SPIE)).

diffraction IL method [8]. The period of the fringe pattern is related to the mother gratings that generate the fringe pattern [9, 10]. For the two-beam diffraction method, the incident beam is diffracted by each grating in certain angles, θm, given by: sinθm = mλ/Pg, where m is the diffraction order and Pg is grating periodicity. When the two gratings are illuminated with the same beam intensity, in the area where the diffracted beams interfere, the periodicity, P, of the aerial image is given by: P=Pg/2 m. Due to the lower diffraction efficiency of diffraction with higher order, only the 1rd diffraction is used for the usual two-beam interference lithography. Thus, the periodicity of the interference beams is given by: P=Pg/2. For the four-beam diffraction method, the periodicity of the interference beams is given by: P=Pg/√2.

Grating is one of the key parts of the XIL techniques. Dry-etching and lift-off processes are usually used for the grating fabrication process. Smooth and steep Cr or Au line structure can be fabricated by using the dry-etching process. However, a poisonous gas such as Cl_2 is usually used during the dry-etching process. Cr grating with 80-nm line period has been fabricated by this process. And line structures with 40-nm period have been printed in a calixarene

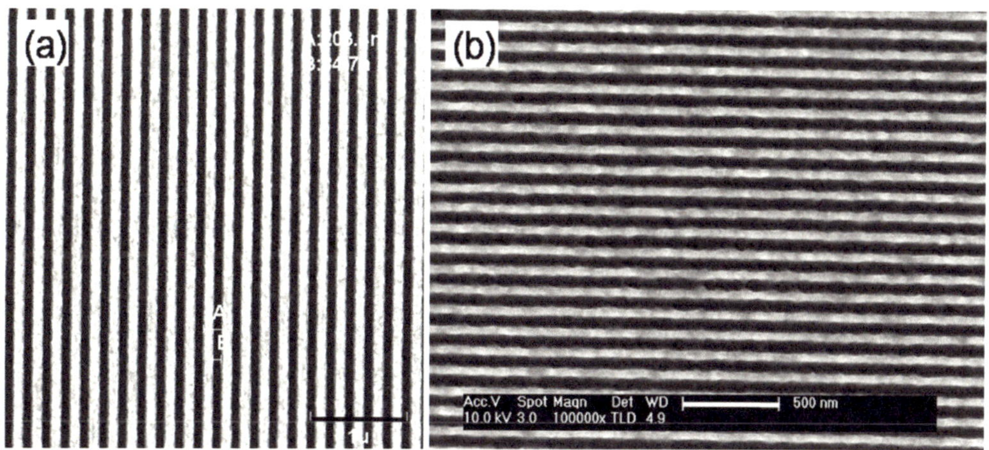

Figure 3. (a) An SEM image of the two beam grating and (b) an SEM image of the exposure result on PMMA resist. The scale bars are 1000 nm in (a) and 500 nm in (b).

negative resist at the XIL II beamline in PSI [11]. By means of electron beam evaporation and lift-off processes, smooth Au line structure can also be fabricated. An Au grating with 200-nm line period has been fabricated by this process, **Figure 3a**. And line structures with 100-nm period have been printed in PMMA resist at the XIL beamline in SSRF, **Figure 3b**.

2. Broad bandwidth IL based on synchrotron radiation source

2.1. BW multi-grating EUV-IL

2D nanopatterns such as s holes, posts, sparse hole arrays or rings can be obtained when the number of interference beams exceeds three. The resultant pattern intensity depends strongly on the relative phases of the beams. Versatile periodic nanostructures can be obtained by changing the number of interfering beams and by controlling the relative phases of the beams [10]. **Figure 4** illustrates the principle of the four-beam interference lithography. The transmission gratings were written on a single mask. The diffracted beams from different gratings overlap to yield a desired pattern. The phases of the diffraction beams are controlled by the precise control of the positions of the gratings, that is, αi, i = 1,⋯, 4, in the figure [10]. Thus, an EBL machine with an interferometer-controlled stage is necessary for the mask-writing process. Two distinct patterns with different contrasts can be obtained in the four-beam case by controlling the phases of the gratings. The incoherent addition of the diffraction beams occurs when $\delta x - \delta y = (n + 1/2)\pi$, where n is an integer and $\delta x = (\alpha 1 - \alpha 2)/2$ and $\delta y = (\alpha 3 - \alpha 4)/2$. The other pattern distribution with high contrast is obtained when $\delta x - \delta y = n\pi$. **Figure 5** illustrates the simulation results demonstrating the influence of phase difference mentioned above during four-beam interference. A four-beam grating with 100 nm is simulated to show

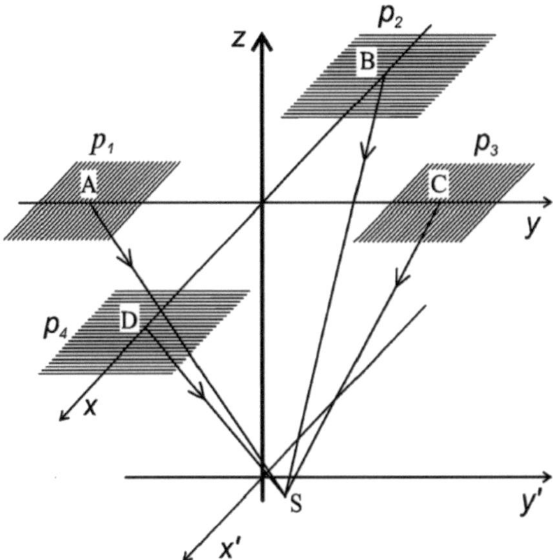

Figure 4. Multiple beam interference lithography with four-beam diffraction gratings (reproduced with permission from [10] ©2005 Elsevier B.V.).

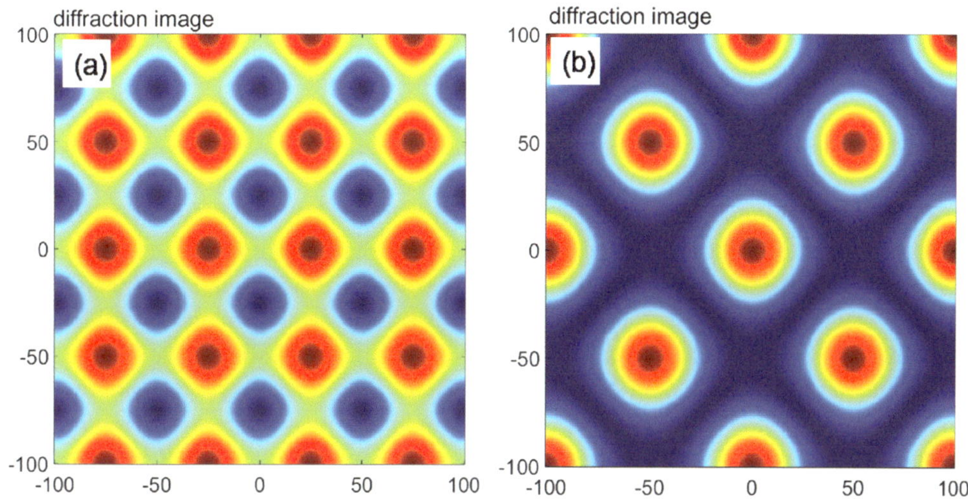

Figure 5. Simulation results demonstrating the influences of phases in four-beam interference. (a) $\delta x - \delta y = (n + 1/2)\pi$, the fringe period in the simulation result is 50 nm. (b) $\delta x - \delta y = n\pi$, the fringe period in the simulation result is 70 nm.

the influence of the phase. Square grids with low-intensity contrast in a 50-nm period (p/2) and a high-intensity contrast in a 70-nm period ($p/\sqrt{2}$) are demonstrated in **Figure 5a** and **5b**.

The exposure results of several masks with different grating arrangements are shown in **Figure 6**. A negative-tone resist, HSQ, was used during exposure. Line-space patterns with 22-nm HP have been obtained by using a typical two-beam grating mask (**Figure 6a**). The exposure of three- and four-beam grating masks will result in hexagonal lattice and square lattice dot arrays in HSQ resist, **Figure 6b** and **6c**. The exposure result of an incoherent illumination mask is shown in **Figure 6d**. Hole arrays with 25-nm HP is obtained in HSQ resist. In this configuration, the two-crossed pair gratings have different periodicities, which resulted in the incoherent addition of the diffraction beams. **Figure 6e** shows a six-grating mask configuration that results in the so-called Kagome lattices [12].

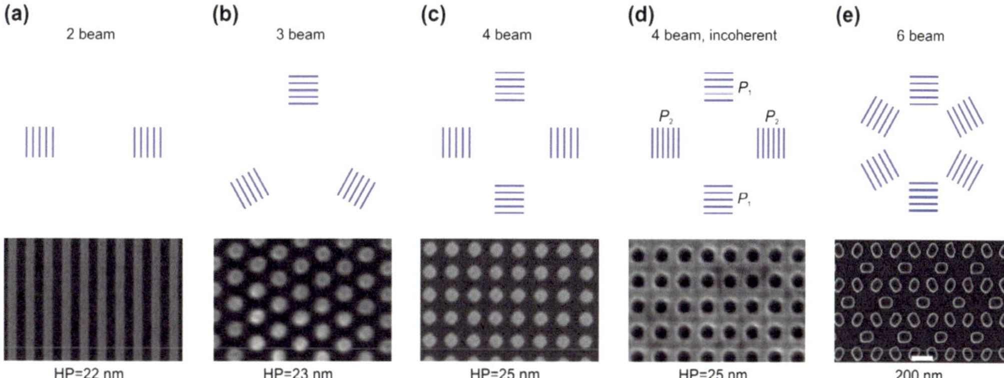

Figure 6. Schematic of different configurations for multiple-beam EUV-IL (first row) and SEM images of corresponding exposures in HSQ photoresist (reproduced with permission from [12] ©2015 Elsevier B.V.).

2.2. Achromatic Talbot lithography

Under monochromatic coherent light illumination, Talbot first noticed that self-images of the gratings were produced at periodic distances away from a transmission diffraction grating [13]. The periodic distance between the self-image planes (Talbot distance, Z_T) is equal to: $Z_T = 2p^2/\lambda$, where p is the grating period and λ is the illumination wavelength. Under the broadband illumination with a spectral bandwidth, the imaging result is very different. Achromatic and stationary patterns of a 1D periodic line structure can be obtained behind the distance $Z_A = 2P^2/\Delta\lambda$ [14]. **Figure 7** shows the scheme of the achromatic and stationary imaging for periodic grating under the broadband illumination with a spectral bandwidth ($\Delta\lambda$). It is called as achromatic spatial frequency multiplication (ASFM) or achromatic Talbot lithography (ATL). The achromatic Talbot distance of the square lattice grid should be $Z_A = 2P^2/\Delta\lambda$, while the achromatic Talbot distance of the hexagonal lattice grid should be $Z_A = 3/2P^2/\Delta\lambda$ [15]. ATL has been proved to be a very robust, highly efficient and simple technique to produce highly dense, high-resolution periodic nanostructures down to 15-nm feature size [16, 17].

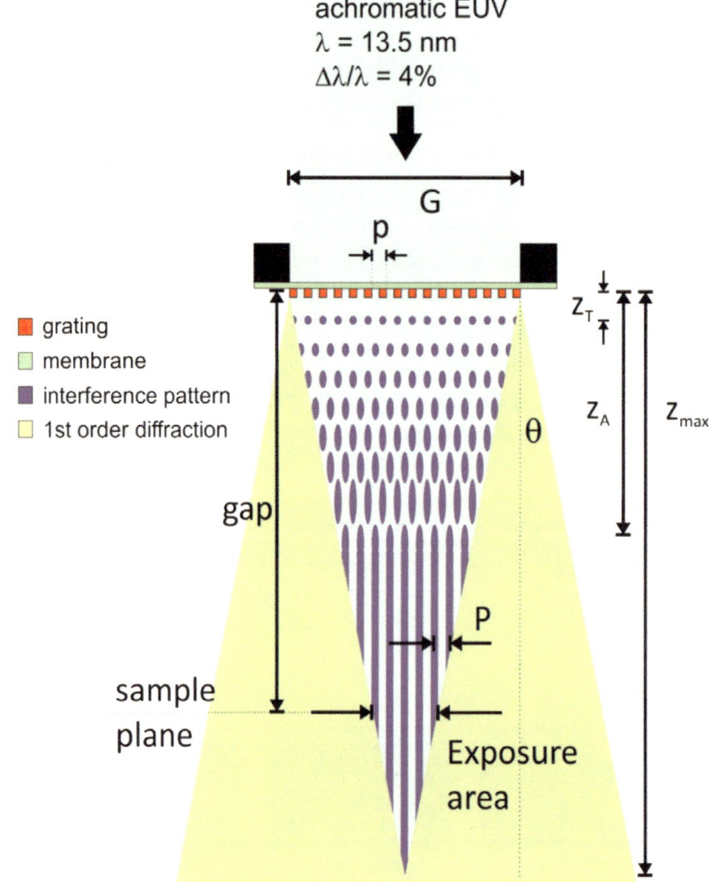

Figure 7. Schematic of the achromatic Talbot lithography. (Reproduced with permission from [16] ©2016 Elsevier B.V.).

http://dx.doi.org/10.5772/intechopen.74564

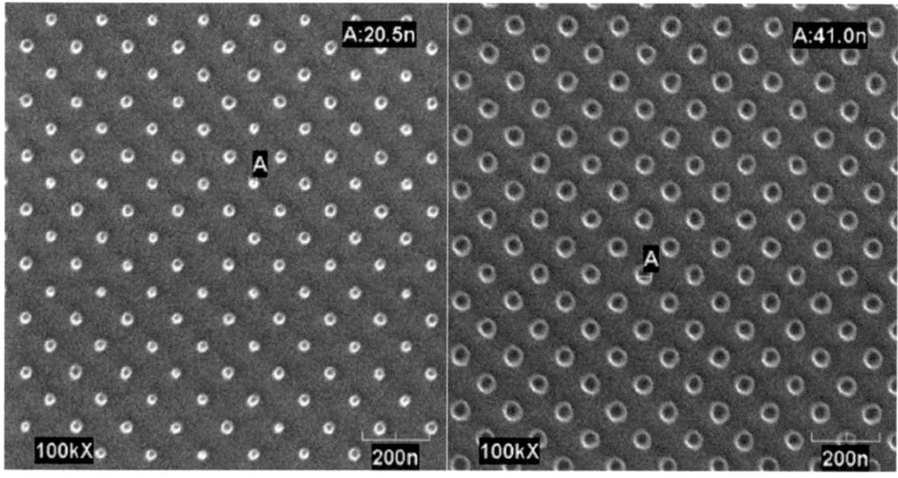

Figure 8. SEM images of the nanodot arrays with dot size at 20 and 40 nm on HSQ resist with a 106-nm period.

During ATL, all of the diffraction orders from different wavelengths overlap together, which makes full use of the beam power. This method is suitable for broadband EUV sources, that is, the majority of EUV sources, and for low-intensity or brightness sources. Two-dimensional periodic patterns in transmission masks with high resolution and uniformity are required to produce dot/hole arrays with high resolution and uniformity. The rectangular transmission grating is very important for the exposure results with high contrast. The nickel or gold electroplating process is utilized to obtain rectangular transmission grating. By using the rectangular transmission grating produced by the nickel electroplating process, nanodot arrays with dot size at 20 and 40 nm in the 106-nm period have been obtained on HSQ resist, as shown in **Figure 8**.

The interference patterns in ATL exposure are much sharper than that in multi-grating IL because there are much more waves with different wave vectors. ATL technique has a long focal length, as well as that of multi-grating one, because of broadband EUV/soft X-rays. However, ATL mask is more difficult to be fabricated than the other since the 0th order light must be blocked completely by the mask.

3. Large-area stitching EUV-IL

Large-area periodic nanostructures are required in many scientific research areas such as nano-magnetics, nano-optics, nano-device fabrication, industrial applications and so on. As mentioned above, large-area periodic nanostructures can be obtained by a single XIL exposure, but the patterned area is limited by the mask area. The area with nanostructures is equal to or less than the area of one grating in the mask. Usually, the grating area is about ~100 × 100 μm^2. This is not large enough for some researches because the spot size of some detecting instruments is already several millimeters. Furthermore, this area is also too small to fabricate a practical device. Unfortunately, it is difficult to stitch the exposure area one by one because the patterned area is surrounded by the area exposed by the 0th order diffraction beams from the mask.

In order to obtain larger exposure-area nanostructures, two scan-exposure techniques have been developed, one is the method employing the achromatic Talbot lithography [17] and the other is based on the BW multi-grating EUV-IL with an order sorting aperture (OSA) [18, 19].

3.1. Step-and-repeat ATL

Compared with EUV-IL, the exposure area of ATL is very similar to that of the mask area, with only small no-interfering areas left on the four sides of the pattern area [17]. To obtain periodic nanopatterns over a large area, step-and-repeat strategy is utilized during ATL. With step-and-repeat ATL, uniform nanopatterns with high uniformity can be produced in the dimension of fabrication capabilities [17]. As mentioned above, ATL makes full use of the beam power. A few seconds is required during the single-shot ATL exposure over an area of about 500 × 500 μm^2. Using step-and-repeat exposure, 15-nm dot arrays over an area of 1 × 1 cm^2 were obtained in just about 5 min by stitching multiple fields [17]. Line arrays up to 5 × 5 mm^2 have also been fabricated by the beam scanning techniques [20]. Comparing to the beam scanning techniques, ATL with step-and-repeat exposure has its advantages. Mask fabrication above the main writing field size of EBL often has challenges and this exacerbates for 2D patterns such as hole or dot arrays. While during ATL with step-and-repeat method, the full use of the beam power make sure the short exposure time for a single-shot exposure over large area. And the large sinle-shot exposure area with small no interfering areas on the four sides enable the easily step-and repeat stitching of multiple exposure fields. This method enables the fabrications of nanopatterns with high resolution and high throughput over large areas.

3.2. Stitching multi-grating EUV-IL

In an EUV-IL, the 0th order diffraction beams from the mask make stitching of the single-shot exposure area impossible. An OSA and an in-situ alignment system are applied to solve this problem in the XIL beamline at SSRF [18]. To block the 0th order diffraction beams through the mask, the OSA size is larger than the pattern area but less than the distance between the grating pairs, as shown in **Figure 9**. The in-situ alignment system contains two parts: a one-dimensional motion motor which is used along the Z direction to adjust the distance between the OSA and the wafer and a two-dimensional motion stage in the XY direction used to align the position of OSA with respect to the mask. By applying this OSA in-situ alignment system, the 0th order diffraction beams could be blocked effectively and the exposure area could be stitched one by one.

Figure 10 shows the stitching result of a four-beam transmission mask with 170-nm period. The 2D nanostructures are exposed over an exposure area of 2 cm x 2 cm by this stitching method, **Figure 10a**. Stitching gaps with no patterns are also shown in **Figure 10b** with less area. It has

http://dx.doi.org/10.5772/intechopen.74564

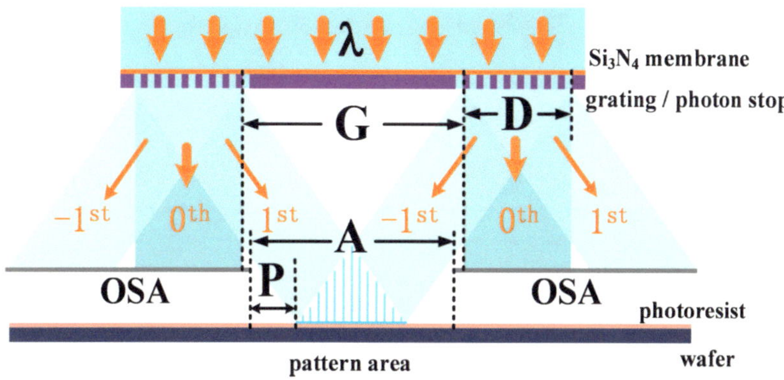

Figure 9. OSA schematic diagram.

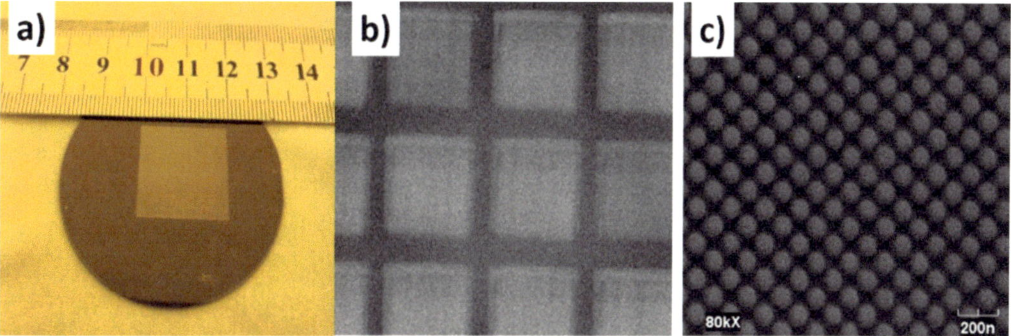

Figure 10. Large-area stitching exposure result: (a) 2 cm × 2 cm total exposure area; (b) stitched exposure blocks; (c) 120-nm period 2D nanostructures in a single block (reproduced from [18], with the permission of AIP publishing).

little impact for some applications for its smaller portion to the total exposure area. These 2D nanostructures show good uniformity with the period of 120 nm, as shown in **Figure 10c**.

4. Soft X-ray IL based on SR

At an EUV photon energy of 92.5 eV, it is difficult to obtain nanopatterns with high aspect ratio due to the strong absorption of conventional photoresist. Normally, the aspect ratio of usual nanopatterns done by EUV-IL is at most ~2 [21]. To improve the aspect ratio of the exposed nanopatterns, soft X-ray interference lithography (SXIL) with higher photon energy is recognized as a better method due to the higher transmission rate in photoresist. Soft X-ray interference lithography with higher photon energies, such as 190, 250 and 450, has already been employed for nanopattern fabrication with higher aspect ratios [12, 22]. High aspect ratio nanopatterns' exposure with high resolution and high uniformity are affected by the incident photon energy, the quality of the mask and the stability of the lithography system. Higher photon energy may decrease the whole optical efficiency of the system, resulting in

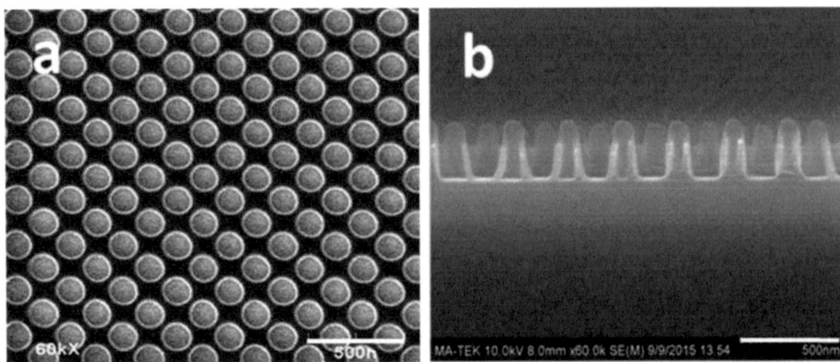

Figure 11. The SEM image of the SXIL exposure result (reproduced with permission from [23] ©2016 Elsevier B.V.).

lower photon flux. A high-absorbed layer is necessary during the exposure with higher photon energy, which makes the masks more fragile. To satisfy a high transmission rate in PMMA with balancing consideration of high photon flux, soft X-ray interference lithography (SXIL) at an energy of 140 eV was carried out at the XIL beamline (BL08U1B) at SSRF [23]. The nanohole array with 200-nm period was successfully obtained in a 300-nm thickness positive PMMA resist, **Figure 11a**. The aspect ratio of the pattern can be up to 3, **Figure 11b**. Compared with the EUV-IL, SXIL can provide a bigger process window for the latter pattern transfer process.

5. Applications

5.1. EUV photoresist evaluation

EUVL with a wavelength of 13.5 nm is thought to be the leading candidate for future nodes of high-volume semiconductor manufacturing. The most important challenges in EUVL include the EUVL system, mask and photoresist. ASML has made great progress toward the high-volume production of the EUVL system. The EUVL system, NXE:3350, is used to expose 1368 wafers per day. ASML expects the first IC manufacturers to start using EUV for chip production from 2018. EUV mask is an integral part of EUV lithography. Intel has installed its pilot production line for EUV mask manufacturing, fixing and detecting. In photolithography, a single defect ruins the chip. Defects on the EUV mask should be detected very clearly. Defects on the EUV mask can be detected by the scanning coherent diffractive imaging methods or by the high-resolution EUV Fresnel zone plate microscope [24].

Photoresist performance is one of the key parameters affecting the performance of EUVL. Due to the really high price of an EUVL system, it is not so easy to evaluate the new resist material by using the practical and affordable optical system. Available projection tools are usually used by researchers to invest in the performance of new resist. However, the resolution of these tools is limited due to the numerical aperture of the projection optics. Based on the coherent radiation from an undulator source, EUV-IL can be used to study the resolution (half pitch, HP), sensitivity (dose) and line-edge roughness of new EUV resist material. In addition to the main challenge of developing high-power EUV sources, EUV-IL has proved to be the best candidate for high-resolution EUV resist evaluation.

EUV-IL enabled the characterization and development of new EUV resist materials before commercial EUV exposure tools became available. EUV resist has been performed on the EUV-IL beamline at SLS [25] and on the XIL beamline at SSRF [26]. Line structures of new EUV resist material with HP of 12 nm have been fabricated at the EUV-IL beamline at SLS, as shown in **Figure 12**. Line structures of new EUV resist material with LWR of 2 nm have been fabricated at the XIL beamline at SSRF, as shown in **Figure 13**. An outgassing test system has been built on the XIL beamline at SSRF too.

5.2. Nano-science

Surface-enhanced Raman scattering (SERS) [27, 28]: Surface-enhanced Raman scattering has been promisingly used in the field of biosensor fabrication with high sensitivity. The enhanced electromagnetic field which happens at the "hotspots" is intimately associated with the high sensitivity. Hotspots can be increased by increasing the density nanoarrays, tuning the shape of nanoparticles or reducing the nanogap between two nanoparticles. Due to its high density and uniformity, EUV-IL has been used for large-area SERS biosensor fabrication.

Large-scale Au nanodisk arrays have been produced on the XIL beamline at SSRF [27]. Nanohole arrays in a 200-nm period were fabricated on the PMMA resist, followed by the Au electron-beam vapor deposition. R6G as low as 10^{-8} M with an enhancement factor of 10^6 has been detected on the Au nanodisk array. High sensitivity with high reproducibility and stability has been verified on the Au nanodisk arrays SERS-active substrates. A total of 32 spot SERS spectra were also collected on the Au nanodisk arrays. The values of RSD of vibrations 1313, 1366 and 1512 cm^{-1} are 18.1, 15.5 and 13.4%. Due to its high density and uniformity, XIL nanofabrication appears to be a promising method for SERS-active substrates' fabrication with high sensitivity and reproducibility.

Metal plasmonic nanostructures with sub-10-nm channels have been fabricated on the EUV-IL beamline at SLS [28]. The SERS signal can be increased by reducing the nanogap

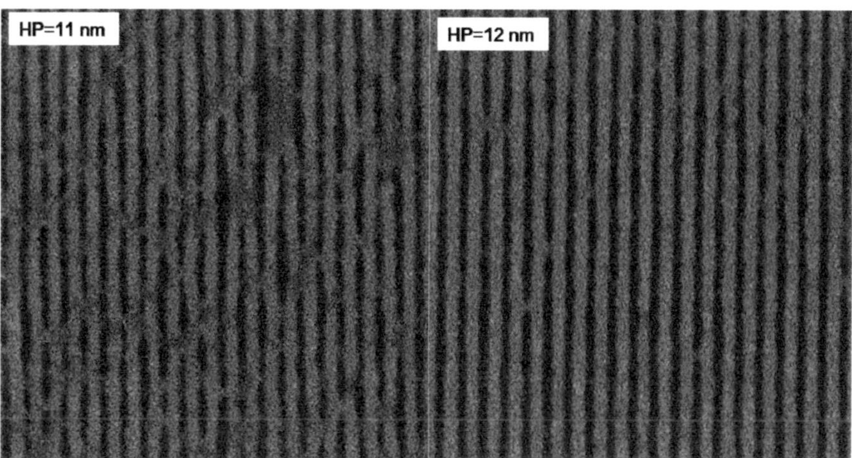

Figure 12. The EUV resist test result, done by XIL-II at SLS (reproduced with permission from [25] ©(2015) COPYRIGHT Society of Photo-Optical Instrumentation Engineers (SPIE)).

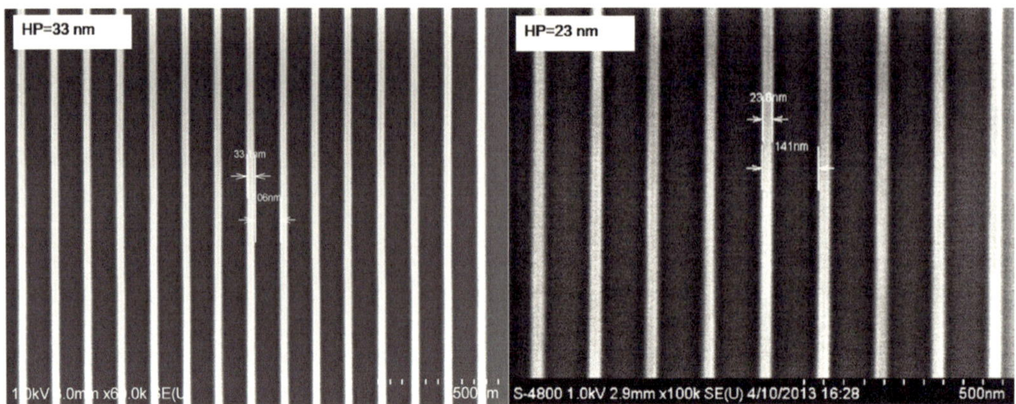

Figure 13. EUV resist test result done by XIL at SSRF.

between two nanoparticles. Double-layer plasmonic nanostructures were fabricated by depositing metal normally onto patterned photoresist layers, which were exposed at the EUV-IL beamline. Metal plasmonic nanostructures with sub-10-nm channels could be produced if the metal layer extended above the photoresist layer, as shown in **Figure 14**. By a comparison with the single-layer antenna, the Raman scattering signal of the double layers with sub-10-nm channels could be improved by a factor of 60. Period nanostructures with high resolution can be done by EUV-IL over a large area at a low cost. Followed by the extended metal layer deposition, high-sensitive SERS-active substrates can be obtained with low cost over large areas, which will be applicable in the near future.

Enhanced light extraction of scintillator [29]: Scintillators are usually used in the radiation detection system. Luminescence in ultraviolet or visible emission can be excited on scintillators by radiation from X-ray, c-ray, electrons, protons and neutrons. Periodic nanostructures over an area of 5.6 × 5.6 mm^2 have been fabricated on the surface of the Bi4Ge3O12 (BGO) scintillator

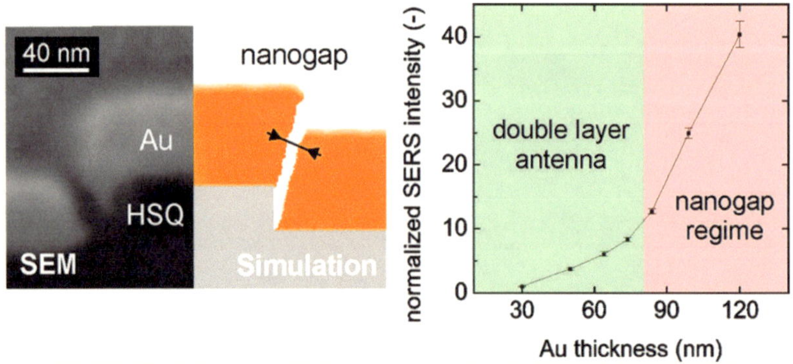

Figure 14. Ballistic simulation of the evaporated cross-section for a double-layer pattern with a nanogap channel (reproduced with permission from [28] ©2014 American Chemical Society).

by the four-beam stitching EUV-IL technique [18]. Combined with conformal deposition of TiO2 by atomic layer deposition, photonic crystal structures over a large area are fabricated. The emission spectra for this photonic crystal with different emergence angles under the excitation at 360 nm have been analyzed. An enhancement factor of 95.1% has been achieved. Photonic crystal structures with large-area and high-index contrast have been fabricated by this method.

Large-area plasmonic color filters [30]: Color filters based on plasmonic nanostructures have received prominent attention in recent years due to their capability of controlling the intensity, phase and polarization of light. Conventional lithography techniques such as electron beam lithography (EBL) and focused ion beam (FIB) are usually used for the plasmonic color filter fabrication. However, the small exposure area and the little throughput of these methods restrict their applications. By the four-beam stitching XIL technique, periodic hole arrays over large areas can be easily fabricated [18]. Followed by the e-beam evaporation of Ag, plasmonic color filters can be produced over large areas. The fill factor of nanostructures can be controlled by changing the dose of exposure. By changing the fill factor, the color can be controlled flexibly. **Figure 15** shows a blue filter device over a large scale fabricated by this method.

High-resolution Fresnel zone plate fabrication [31]: High-resolution Fresnel zone plates (FZPs) are often used in high-resolution x-ray microscopy. The resolution of the x-ray microscopy is limited to its outermost zone of the FZP which is used as a lens. EBL is usually used for the

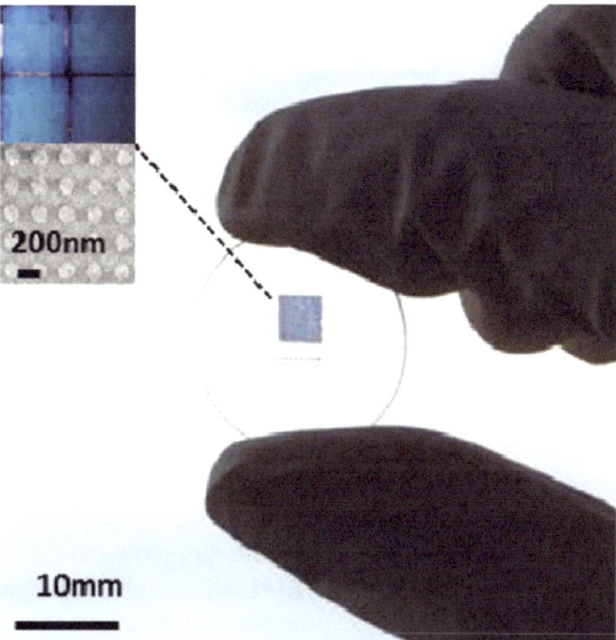

Figure 15. One blue filter device fabricated by stitching XIL, held by fingers (reproduced with permission from [30] ©2016 Optical Society of America).

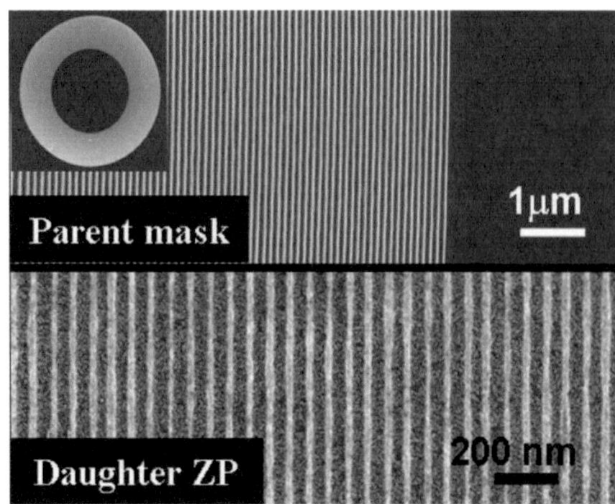

Figure 16. SEM images of the outer zones of a parent mask and the corresponding daughter ZP are shown in the SEM images. Inset shows the full ZP on the mask. (reproduced with permission from [31] ©2011 Optical Society of America).

FZP fabrication with high resolution. Based on the self-imaging (ASFM) property of gratings, high-resolution Fresnel zone plates (FZPs) have been fabricated on the EUV-Il beamline in PSI. ASFM is also known as ATL. Under wide-band illumination with spectral width $\Delta\lambda$, achromatic and stationary imaging of the original grating can be obtained beyond the distance $Z_A = 2P^2/\Delta\lambda$. Under broadband EUV illumination, a radially oscillating intensity distribution with double the spatial frequency of the parent ZP is produced. This intensity distribution is observed in a certain distance range, which can be used to record daughter ZPs with half the zone width of the parent ZPs. FZPs with zone widths as low as 30 nm have been fabricated, as shown in **Figure 16**. FZP fabrication with high resolution and high throughput can be fabricated in this way.

6. Conclusion(s)

Extreme UV interference lithography (EUV-IL) is a useful tool to fabricate periodic nanostructures and is considered as an ideal method for EUV photoresist evaluation. Synchrotron radiation beamlines provide stable sources with full spatial coherence, which are enough to support EUV-IL and necessary to further support soft X-ray interference lithography (SXIL) with higher photon energies. The diffraction schemes are limited to achromatic ones (SXIL) in usual multi-grating interference lithography because the synchrotron radiation beams have limited bandwidths. However, the limited bandwidth is essential to be employed in achromatic Talbot lithography (ATL) with a large focal length. Based on the two interference lithography methods, step-and-repeat ATL and stitching Multi-grating EUV-IL/SXIL have been developed to fabricate centimeter-scale periodic structures. Many nano-science applications have been illustrated, with EUV-IL as fabrication tools.

Acknowledgements

This work was supported by National Key R&D Program of China (2016YFA0401302),the National Natural Science Foundation of China (No. 11775291,11505275, 11475251,11275255), the Youth Innovation Promotion Association (No. 2017306), the Open Research Project of Large Scientific Facility from Chinese Academy of Sciences: Study on Self-Assembly Technology and Nanometer Array with Ultrahigh Density. The authors thank the support of BL08U1B and BL08U1A at SSRF for sample preparation. The authors also thank Dr. H.H. Solak (EULITHA AG) and Dr. Y. Ekinci (XIL-II, SLS) for helpful discussions on EUV-IL and ATL. Parts of this chapter are taken from the authors' former work at the SSRF beamline BL08U1B [2], on ATL [15, 16], SXIL [23] and stitching multi-grating EUV-IL/SXIL [17, 18] and parts of the nano-science applications of EUV-IL are taken from [27–31].

Conflict of interest

I confirm there are no conflicts of interest.

Author details

Shumin Yang and Yanqing Wu*

*Address all correspondence to: wuyanqing@sinap.ac.cn

Shanghai Institute of Applied Physics, Shanghai Synchrotron Radiation Facility, Shanghai, China

References

[1] Buitrago E, Fallica R, Fan D, Karim W, Vockenhuber M, van Bokhoven AJ, Ekinci Y. From powerful research platform for industrial EUV photoresist development, to world record resolution by photolithography: EUV interference lithography at the Paul Scherrer Institute. Proceedings of SPIE. 2016;**9426**:94260T-12. DOI: 10.1117/12.2238805

[2] Yang SM, Wang LS, Zhao J, Xue CF, Liu HG, Xu ZJ, Wu YQ, Tai RZ. Developments at SSRF in soft X-ray interference lithography. Nuclear Science and Techniques. 2015;**26**:010101-010107. DOI: 10.13538/j.1001-8042/nst.26.010101

[3] Fukushima Y, Sakagami N, Kimura T, Kamaji Y, Iguchi T, Yamaguchi Y, Tada M, Harada T, Watanabe T and Kinoshita H. Development of extreme ultraviolet interference lithography system. Japanese Journal of Applied Physics. 2010;**49**:06GD06-06GD05. DOI: 10.1143/JJAP.49.06GD06

[4] Isoyan A, Wüest A, Wallace J, Jiang F, Cerrina F. 4X reduction extreme ultraviolet interferometric lithography. Optics Express. 2008;**16**:9106-9111. DOI: 10.1364/OE.16.009106

[5] Lin CH, Fong CH, Lin YM, Lee YY, Fung HS, Shew BY, Shieh J. EUV interferometric lithography and structural characterization of an EUV diffraction grating with nondestructive spectroscopic ellipsometry. Microelectronic Engineering. 2011;**88**:2639-2643. DOI: 10.1016/j.mee.2011.02.002

[6] Solak HH, He D, Li W, Singh-Gasson S, Cerrina F, Sohn BH, Yang XM, Nealey P. Exposure of 38 nm period grating patterns with extreme ultraviolet interferometric lithography. Applied Physics Letters. 1999;**75**:2328-2330. DOI: 10.1063/1.125005

[7] Solak HH, Li W, He D, Wallace J, Cerrina F. A new beamline for EUV lithography research. AIP Conference Proceedings. 2000;**521**:99-103. DOI: 10.1063/1.1291766

[8] Auzelyte V, Dais C, Farquet P, Grützmacher D, Heyderman LJ, Luo F, Olliges S, Padeste C, Sahoo PK, Thomson T, Turchanin A, David C, Solak HH. Extreme ultraviolet interference lithography at the Paul Scherrer Institut. Journal of Micro/Nanolithography, MEMS, and MOEMS. 2009;**8**(2):021204-0212010. DOI: 10.1117/1.3116559

[9] Solak HH, David C, Gobrecht J, Wang L, Cerrina F. Multiple-beam interference lithography with electron beam written gratings. Journal of Vacuum Science & Technology B: Microelectronics and Nanometer Structures Processing, Measurement, and Phenomena. 2002;**20**(6):2844-2848. DOI: 10.1116/1.1518015

[10] Solak HH. Space-invariant multiple-beam achromatic EUV interference lithography. Microelectronic Engineering. 2005;**78-79**:410-416. DOI: 10.1016/j.mee.2005.01.012

[11] Solak HH. Nanolithography with coherent extreme ultraviolet light. Journal of Physics D: Applied Physics. 2006;**39**:R171-R188. DOI: 10.1088/0022-3727/39/10/R01

[12] Mojarad N, Gobrecht J, Ekinci Y. Interference lithography at EUV and soft X-ray wavelengths: Principles, methods, and applications. Microelectronic Engineering. 2015;**143**: 55-63. DOI: 10.1016/j.mee.2015.03.047

[13] Talbot HF. Facts relating to optical science No. IV. The London, Edinburgh, and Dublin Philosophical Magazine and Journal of Science. 1836;**9**(56):403-407. DOI: 10.1080/147864 43608649032

[14] Guérineau N, Harchaoui B, Primot J. Talbot experiment re-examined: Demonstration of an achromatic and continuous self-imaging regime. Optics Communications. 2000;**180**:199-203. DOI: 10.1016/S0030-4018(00)00717-3

[15] Yang SM, Zhao J, Wang LS, Zhu FY, Xue CF, Liu HG, Wu YQ, Tai RZ. Influence of symmetry and duty cycles on the pattern generation in achromatic Talbot lithography. Journal of Vacuum Science & Technology B. 2017;**35**(2):021601-021607. DOI: 10.1116/1.4974930

[16] Fan D, Buitrago E, Yang SM, Karim W, Wu YQ, Tai RZ, Ekinci Y. Patterning of nanodot-arrays using EUV achromatic Talbot lithography at the Swiss Light Source and Shanghai Synchrotron Radiation Facility. Microelectronic Engineering. 2016;**155**:55-60. DOI: 10.1016/j.mee.2016.02.026

[17] Karim W, Tschupp SA, Oezaslan M, Schmidt JT, Gobrecht J, Bokhoven JA, Ekinci Y. High-resolution and large-area nanoparticle arrays using EUV interference lithography. Nanoscale 2015;**7**:7386-7393. DOI: 10.1039/c5nr00565e

[18] Xue CF, Wu YQ, Zhu FY, Yang SM, Liu HG, Zhao J, Wang LS, Tai RZ. Development of broadband X-ray interference lithography large area exposure system. Review of Scientific Instruments. 2016;**87**:043303-043304. DOI: 10.1063/1.4947067

[19] Xue C, Zhao J, Wu YQ, Yu HN, Yang SM, Wang LS, Zhao WC, Wu Q, Zhu ZC, Liu B, Zhang X, Zhou WC, Tai RZ. Fabrication of large-area high-aspect-ratio periodic nanostructures on various substrates by soft X-ray interference lithography. Applied Surface Science. 2017;**425**:553-557. DOI: 10.1016/j.apsusc.2017.07.010

[20] Wang L, Solak HH, Ekinci Y. Fabrication of high-resolution large-area patterns using EUV interference lithography in a scan exposure mode. Nanotechnology 2012;**23**:305303-305305. DOI: 10.1088/0957-4484/23/30/305303

[21] Solak HH, David C, Gobrecht J, Wang L, Cerrina F, Four-wave E. UV interference lithography. Microelectronic Engineering. 2002;**61-62**:77-82. DOI: 10.1016/S0167-9317(02)00579-8

[22] Mojarad N, Fan D, Gobrecht J, Ekinci Y. Broadband interference lithography at extreme ultraviolet and soft X-ray wavelengths. Optics Letters. 2014;**39**(8):2286-2289. DOI: 10.1364/OL.39.002286

[23] Zhao J, Wu YQ, Xue CF, Yang SM, Wang LS, Zhu FY, Zhu ZC, Liu B, Wang Y, Tai RZ. Fabrication of high aspect ratio nanoscale periodic structures by the soft X-ray interference lithography. Microelectronic Engineering. 2017;**170**:49-53. DOI: 10.1016/j.mee.2016.12.028

[24] Helfenstein P, Mohacsi I, Rajeev R, Ekinci Y. Scanning coherent diffractive imaging methods for actinic extreme ultraviolet mask metrology. Journal of Micro/Nanolithography, MEMS, and MOEMS. 2016;**15**(3):034006-034005. DOI: 10.1117/JMM.15.3.034006

[25] Buitrago E, Yildirim O, Verspaget C, Tsugama N, Hoefnagels R, Rispens G, Ekinci Y. Evaluation of EUV resist performance using interference lithography. Proceedings of SPIE. 2015;**9422**:94221S-1-94221S-13. DOI: 10.1117/12.2085803

[26] Chen L, Xu J, Yuan H, Yang SM, Wang LS, Wu YQ, Zhao J, Chen M, Liu HG, Li SY, Tai RZ, Wang SQ, Yang GQ. Outgassing analysis of molecular glass photoresists under EUV irradiation. Science China Chemistry. 2014;**57**(12):1746-1750. DOI: 10.1007/s11426-014-5122-y

[27] Zhang PP, Yang SM, Wang LS, Zhao J, Zhu ZC, Liu B, Zhong J, Sun XH. Large-scale uniform Au nanodisk arrays fabricated via X-ray interference lithography for reproducible and sensitive SERS substrate. Nanotechnology. 2014;**25**:245301-245308. DOI: 10.1088/0957-4484/25/24/245301

[28] Siegfried T, Wang L, Ekinci Y, Martin QJF, Sigg H. Metal double layers with Sub-10 nm channels. ACS Nano. 2014;**8**(4):3700-3706. DOI: 10.1021/nn500375z

[29] Zhu ZC, Wu S, Xue CF, Zhao J, Wang LS, Wu YQ, Liu B, Cheng CW, Gu M, Chen H, Tai RZ. Enhanced light extraction of scintillator using large-area photonic crystal structures fabricated by soft-X-ray interference lithography. Applied Physics Letters. 2015;**106**:241901-241905. DOI: 10.1063/1.4922699

[30] Sun LB, Hu XL, Wu QJ, Wang LS, Zhao J, Yang SM, Tai RZ, Fecht HJ, Zhang DX, Wang LQ, Jiang JZ. High throughput fabrication of large-area plasmonic color filters by soft-X-ray interference lithography. Optics Express. 2016;**24**(17):19112-19121. DOI: 10.1364/OE.24.019112

[31] Sarkar SS, Solak HH, Saidani M, David C, Friso van der Veen J. High-resolution Fresnel zone plate fabrication by achromatic spatial frequency multiplication with extreme ultraviolet radiation. Optics Letters. 2011;**36**(10):1860-1862. DOI: 10.1364/OL.36.001860

Chapter 6

Optical Proximity Correction (OPC) Under Immersion Lithography

Ahmed Awad, Atsushi Takahashi and
Chikaaki Kodaman

Additional information is available at the end of the chapter

http://dx.doi.org/10.5772/intechopen.72699

Abstract

As advanced technology nodes continue scaling down into sub-16 nm regime, optical microlithography becomes more vulnerable to process variations. As a result, overall lithographic yield continuously degrades. Since next-generation lithography (NGL) is still not mature enough, the industry relies heavily on resolution enhancement techniques (RETs), wherein optical proximity correction (OPC) with 193 nm immersion lithography is dominant in the foreseeable future. However, OPC algorithms are getting more aggressive. Consequently, complex mask solutions are outputted. Furthermore, this results in long computation time along with mask data volume explosion. In this chapter, recent state-of-the-art OPC algorithms are discussed. Thereafter, the performance of a recently published fast OPC methodology—to generate highly manufactured mask solutions with acceptable pattern fidelity under process variations—is verified on the public benchmarks.

Keywords: immersion lithography, optical proximity correction (OPC), mask, edge placement error (EPE), process variability band (PV band), runtime, mask manufacturability, kernel

1. Introduction

Optical microlithography provides a feasible solution in the foreseeable future for advanced technology nodes patterning with its relatively cheap equipment, if compared with other fabrication techniques. An integrated circuit (IC) design level elements are represented as a set of polygons that are carved onto a pixelated template, called the mask. Mask image is then projected onto a photoresist coating the silicon wafer through an exposure tool. If sufficient light intensity is projected onto the resist, it is chemically exposed. Exposed regions are then etched to form the target circuitry pattern onto the silicon wafer [1, 2].

Complex circuit is made up by repeating lithographic operation for each layer. With the continuous shrinkage of critical dimensions (CDs) of advanced technology nodes following Moore's law, IC dimensions are being pushed into sub-16 nm according to the International Technology Roadmap for Semiconductors (ITRS) [3]. Thus, light diffraction and interference impact becomes pronounced during circuit printing, which results in wafer image quality degradation. For example, corners are rounded and lines are shortened. Such distortions in the wafer image impact circuit functionality and performance. Besides, it could result in circuit malfunction [4, 5].

To reduce the minimum printable CD, wavelength of the illumination source of the optical system had been steadily reduced till it reached its practical limit at 193 nm due to high instability and strong birefringence of lens materials [6]. Immersion lithography has been introduced to improve the resolution through filling the gap between wafer and projection lens with purified water for higher numerical aperture. However, CD of technology nodes continues scaling down to become small fractions of the wavelength. This makes 193 nm immersion lithography insufficient for modern ICs printing [7].

Resolution enhancement techniques (RETs) aim to improve wafer image quality through manipulating the amplitude and phase of the optical wave to pre-compensate wafer image distortions [8]. Since next-generation lithography (NGL) is still not mature enough, the industry relies heavily on RETs, wherein optical proximity correction (OPC) is dominant, to print sub-16 nm technology nodes in the foreseeable future [9].

In OPC, a mask pattern is iteratively adjusted to obtain an acceptable wafer image quality. However, a lithographic process is susceptible to raw process variations, which result in lithographic yield degradation. Since finding an optimal mask solution with acceptable wafer image quality under all possible process conditions is infeasible, the industry defines a process window including a set of process conditions upon request. The most probable process condition is often defined as nominal process condition under which acceptable wafer image quality is desired with minimizing the variations between different images within the process window [10, 11].

To keep pace with advanced technology nodes, model-based OPC algorithms get increasingly more aggressive. Consequently, complex mask solutions are outputted, which results in mask manufacturability degradation along with explosion in mask data volume [12, 14].

OPC computation time forms another crucial factor. For example, brute force algorithms to find optimal mask solutions are infeasible for industrial cases, wherein, mask data have to be prepared in a matter of hours to cover the huge number of target circuitries [10, 13].

In this chapter, recent state-of-the-art OPC algorithms are discussed. Thereafter, a recently published algorithm in [15, 16] is deeply analyzed and its performance is verified in terms of pattern fidelity under process variations, mask manufacturability, and computation time.

The rest of this chapter is organized as follows: Section 2 briefly discusses recent OPC algorithms and their main shortcomings. Section 3 describes lithographic terminology and mask evaluation metrics. Sections 4 and 5 discuss intensity modeling and the to-be-evaluated OPC methodology, respectively. Experimental results are proposed in Section 6, and Section 7 concludes the chapter.

2. Recent research

Several algorithms have been proposed to minimize edge placement error (around wafer image contours) in model-based OPC. Mask error enhancement factor (MEEF) matrix has been widely adopted to guide edge shifting following EPE changes in each fragment control point [17, 18]. However, such algorithms slowly converge for advanced technology nodes. Source and mask optimization has been proposed in [19] at the cost of long computational time. A fast intensity-based algorithm has been proposed in [20]. However, this algorithm considers only sparse patterns, while recent dense patterns are more challenging. Adaptive fragments refinement has been proposed to improve wafer image quality without significantly considering process variations [21].

Process window OPC algorithms consider both EPE and process variations [22, 23]. However, wafer image has to be simulated under each process condition, which is time-consuming.

Retargeting has been adopted to improve pattern fidelity through modifying the target pattern itself along with the mask at the cost of long computation time [24]. Process variations have been effectively considered through including the intensity slope in the cost function in simultaneous mask and target optimization (SMATO) algorithm [25].

Inverse lithography technology (ILT) has been extensively exploited to find optimal mask solutions based on rigorous mathematical models [26, 27]. However, ILT masks are hard to be manufactured due to ILT pixel-based behaviors.

Sub-resolution assist features (SRAFs) insertion has been widely exploited to increase mask robustness against dose variations [28, 29]. Consideration of multiple process conditions is required upon SRAF insertion/sizing.

To improve mask manufacturability without sacrificing lithographic yield, design aware OPC algorithms include a set of restricted design rules (RDRs) in the OPC recipe. RDRs define the minimum dimensions in mask geometry [30, 31]. Although including RDRs in the OPC preserves acceptable pattern fidelity with less complex masks, long computation time is expected due to the low stability and slow convergence of the algorithm.

To accelerate OPC computation, a fast method has been proposed in [32] to simulate wafer image with less number of kernels. However, using more kernels is required in further iterations. Intensity difference map has been recently proposed in [15] and its performance has been confirmed in [33].

Recently, an effective Process Variation Aware OPC algorithm, namely PV-OPC, has been proposed with good results in terms of pattern fidelity under process variations, computational time, and with considering mask notch rule for higher manufacturability through exploiting variational EPE, and adaptive fragmentation [9]. Furthermore, PV-OPC effectively reduces the number of needed simulations. Mask Optimization Solution with Process Window Aware Inverse Correction (MOSAIC) algorithm has been recently proposed as an ILT algorithm with exploiting variational EPE under each process corner. MOSAIC has two versions: fast and exact [34]. However, complex masks are outputted from this algorithm. A robust approach for process variation OPC has been recently published in [35] at the cost of outputting complex masks.

Recently, a novel intensity-based OPC methodology has been published in [15]. This algorithm outperforms the most recent effective algorithms on the public benchmarks in terms of pattern fidelity under process variations and runtime. Besides, this algorithm has been extended to improve mask manufacturability in [16] with preserving its effectiveness. This algorithm is analyzed in this chapter and its effectiveness is numerically verified through comparing it with other recent algorithms on the most challenging public benchmarks.

3. Lithographic terminology and problem description

A lithographic process is susceptible to raw process variations resulting in lithographic yield degradation. Dose and focus variations are dominant in this context. Thus, a set of dose and focus process conditions that are requested to be considered are defined as a process window P_w.

3.1. Lithographic pattern terminology

Given a region of pixels R wherein a target pattern T is defined such that $T \subset R$. Similarly, a mask pattern M is defined in R such that $M \subset R$.

A pattern consists of a set of nonoverlapped rectilinear polygons where a polygon consists of a set of connected pixels. Let S be a polygon. If a pixel p is contained in S, it is denoted by $p \in S$. Furthermore, if $p \in S \in T$, it is denoted by $p \in T$. The same notation is applied for a pixel $p \in S \in M$, which is simply denoted by $p \in M$.

An edge on the boundary of a polygon is either a horizontal or vertical line connecting two corners. Let E_T and E_M be the set of edges along the boundary of all polygons in T and M, respectively. Let $l(e)$ denote the length of an edge e and $D(e_i, e_j)$ denote the Manhattan distance between edges e_i and e_j in a target/mask pattern.

Target pattern: A set of target design rules are defined to be satisfied. This includes: (1) minimum allowable line width, denoted by L_w. (2) Minimum spacing between different polygons, denoted by L_s. Note that, $e \in E_T; l(e) \geq L_w$.

A corner on the boundary of a polygon in the target is either positive or negative. A positive corner forms 90° angle outside the polygon, while a negative corner forms 270° angle inside the polygon. **Figure 1(a)** illustrates a target pattern with both types of corners [15].

Mask pattern: Mask pattern polygons are classified into three types: core-polygon, serif, and SRAF. **Figure 1(b)** shows these types. A core-polygon that corresponds to a polygon $S \in T$ is obtained from S by fragmenting its boundary to segments and shifting them. A segment located on a corner in the target pattern is said to be a corner segment. A serif on a positive corner is a squared feature added outside of the polygon, while, a serif on a negative corner is a square picked from the polygon. An SRAF (scatter bar) is a long bar parallel to an edge of a polygon in the target [15].

A notch is either peak or valley in the polygon geometry, as illustrated in **Figure 2(a)** [36]. From mask manufacturing perspective, thin notches are forbidden. A jog is the orthogonal edge between two neighboring edges in the mask boundary. Small jogs in an OPC mask typically

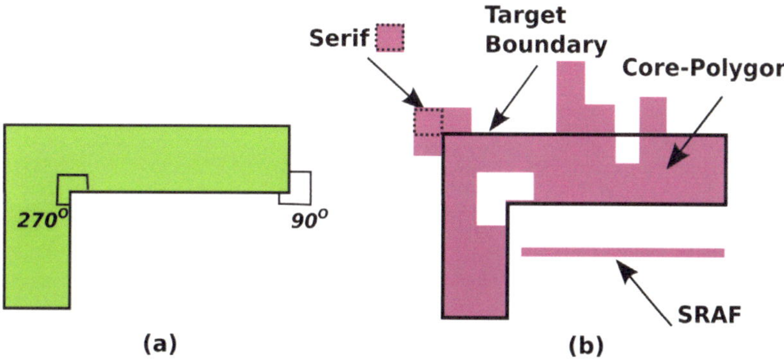

Figure 1. (a) Target pattern T. (b) Mask pattern M [15].

exist, as shown in **Figure 2(b)**. However, such small features increase shot count during mask writing [12]. Moreover, they increase mask manufacturing process variations, which turns out into pattern fidelity degradation.

Mask design rules define a set of constraints to be satisfied in a mask pattern for higher mask manufacturability. In this chapter, the following mask design rules are considered: (1) mask notch rule, which defines the minimum allowable edge length in the mask boundary, denoted by d_n. (2) Mask spacing rule, which defines the minimum allowable spacing between patterns in the mask pattern, denoted by d_s.

3.2. Lithographic model

A mask M is transformed through an optical and projection system into an aerial image. This image is an intensity map holding a set of light intensities floating onto the resist. The set of exposed pixels within the intensity map forms the image onto the silicon wafer. Let $I_{P_c}(M)$ and $G_{P_c}(M)$ represent the intensity map and wafer image of mask M under process condition $P_c \in P_w$, respectively, as illustrated in **Figure 3**.

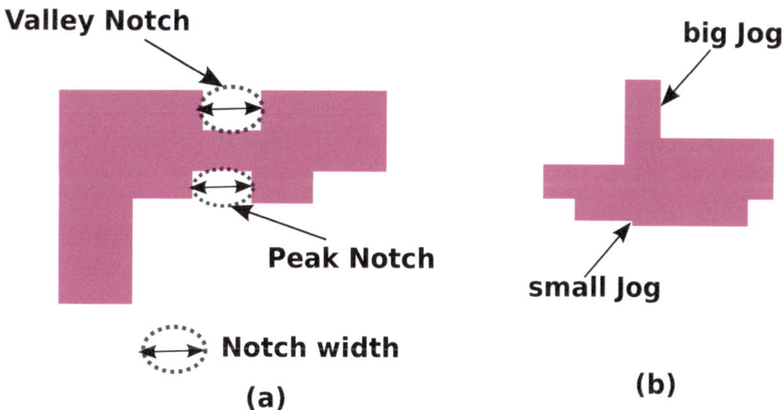

Figure 2. (a) Notch types and width and (b) jogs in the mask boundary.

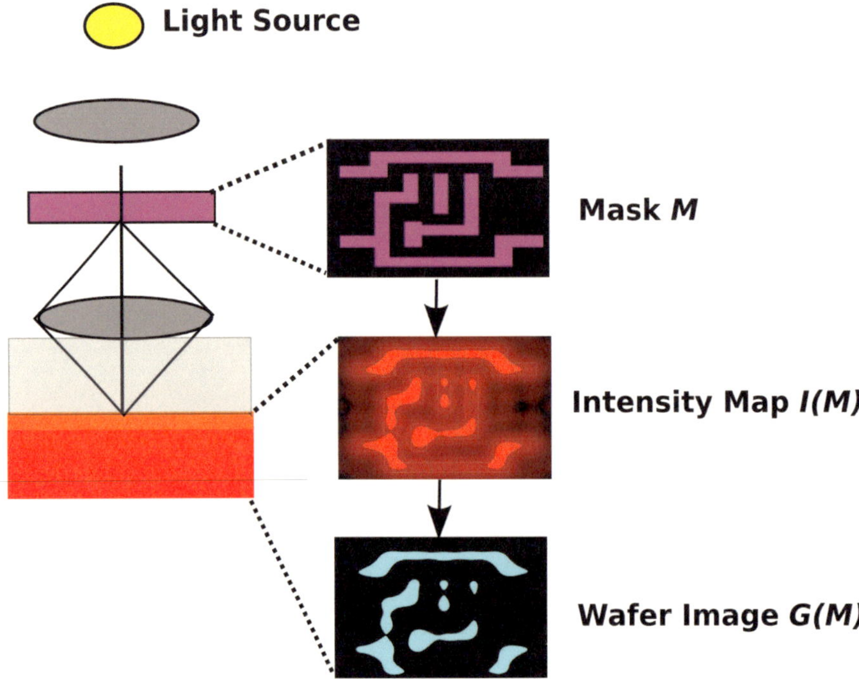

Figure 3. Intensity map and wafer image for a given mask pattern.

Sum of coherent systems (SOCS) is often used in OPC to roughly estimate the intensity map [38]. In SOCS model, the optical system is decomposed into a set of coherent kernels working as low pass filters. Each kernel has an eigenfunction, which represents its filtering behavior and eigenvalue and its weight for intensity estimation. For a mask M, intensity map, $I_{Pc}(M, K)$, under process condition P_c is defined as given in Eq. (1), where K denotes the set of all kernels in a lithographic system, $\sigma_k{}^{P_c}$ and $\phi_k^{P_c}$ represent the eigenvalue and the eigenfunction for kernels $k \in K$ under process condition P_c, respectively, and $\otimes$ denotes convolution operation.

$$I_{P_c}(M) = \sum\nolimits_{k \in K} \sigma_k{}^{P_c} \left| \phi_k^{P_c} \otimes M \right|^2 \tag{1}$$

Once intensity map is obtained, it undergoes resist modeling. Constant threshold resist (CTR) is one of the commonly used resist models, wherein intensity threshold of exposure I_{th} is predefined. Wafer image $G_{Pc}(M)$ is the set of pixels whose intensities are greater than or equal to I_{th}, as given in Eq. (2), where $I_{Pc}(M, p)$ represents the intensity in pixel p by mask M.

$$G_{P_c}(M) = \{p \in R | I_{P_c}(M) \geq I_{th}\} \tag{2}$$

3.3. Representative lithographic process conditions

Wafer image gets wider with higher positive dose values. On the other hand, it gets thinner with negative values. Defocus impact causes wafer image to be thinner than its form under

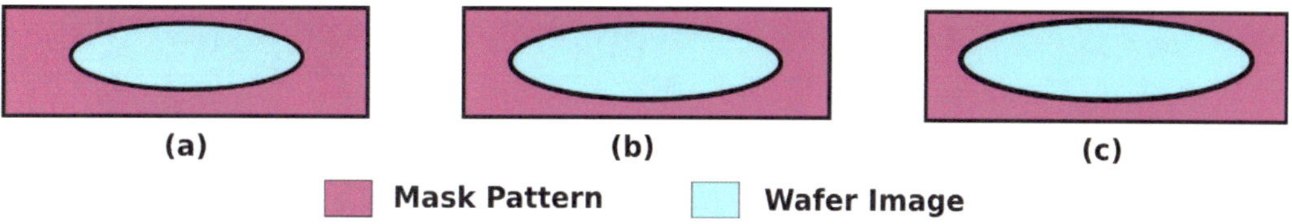

Figure 4. Representative wafer images: (a) innermost, (b) nominal, (c) outermost.

nominal focus condition [9]. Thus, for a process window P_w, three representative process conditions are defined as follows (illustrated in **Figure 4**):

1. Innermost process condition: Includes the maximum negative dose value and defocus under which innermost intensity map $I_i(M)$, is defined. Innermost wafer image $G_i(M)$ is extracted from $I_i(M)$.

2. Outermost process condition: Includes the maximum positive dose and in-focus under which maximum intensity map $I_o(M)$, is defined. Outermost wafer image $G_o(M)$ is extracted from $I_o(M)$.

3. Nominal process condition: Includes average dose and in-focus under which nominal intensity map $I_n(M)$, is defined. Nominal wafer image $G_n(M)$ is extracted from $\mathrm{I_n(M)}$.

3.4. Mask evaluation metrics

A mask pattern is evaluated in terms of the pattern fidelity under nominal process condition, robustness against process variations, mask manufacturability, and the computation time required to find that mask solution.

Pattern fidelity evaluation: Edge placement error (EPE) is often used for pattern fidelity evaluation under nominal process condition. EPE is the geometrical distance between a point on the target boundary and its corresponding point onto wafer image contour. Let $epe(M, t)$ denote the EPE for a point $t \in T$, as shown in **Figure 5**. As long as no electric violations occur in the circuit functionality, EPE evaluation can be relaxed. Let EPE_{max} be the maximum allowable EPE distance [10].

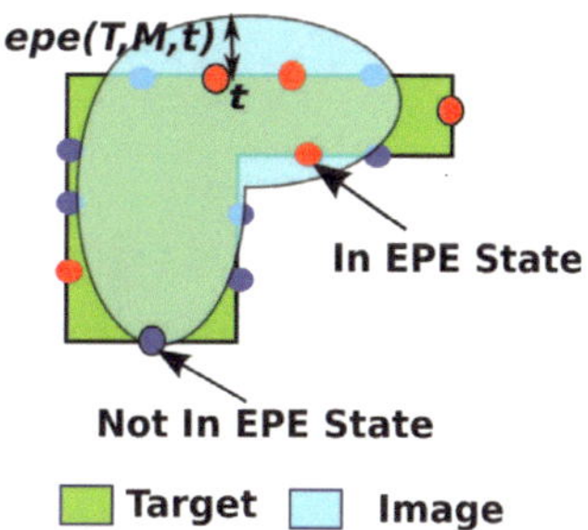

Figure 5. EPE evaluation [15].

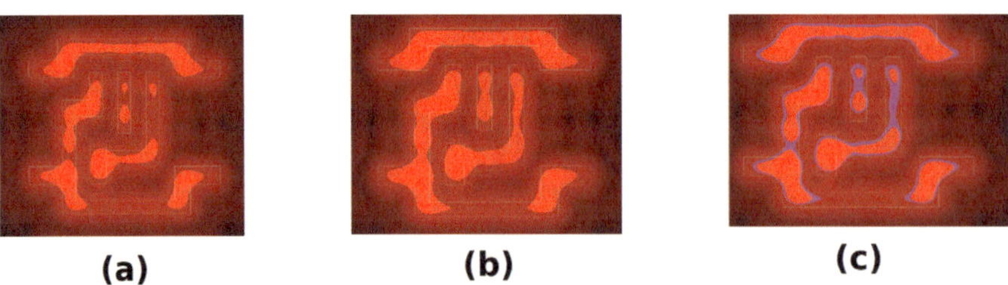

Figure 6. (a) Innermost intensity map, (b) outermost intensity map and (c) PV band area.

For fast evaluation, EPE is statically measured among a set of tap points defined on the boundary of *T*, as given in **Figure 5**. Let *A* denote the set of defined tap points. For each $t \in A$, let $t^+ \in T$ be a point whose distance from *t* is EPE_{max} pixels, which is on the line that passes *t* and perpendicular to its edge in the target. Similarly, $t^- \in T$ is defined but inside the polygon in *T*. For a tap point $t \in A$, it is said to be not in EPE state if $I_n(M, t^-) \geq I_{th}$ and $I_n(M; t^+) < I_{th}$. Otherwise, *t* is said to be in EPE state. The number of EPE violations for mask M, denoted by *#EPEV(M)*, is the number of tap points in EPE state. Pattern fidelity of a mask *M* is assumed to be inversely proportional to *#EPEV(M)* [15].

Process variability evaluation: Process variability (PV) band area is a commonly used metric for process variations. PV band area is the area denoted by XORing wafer images under all process conditions within process window P_w.

Innermost and outermost wafer images are exploited to provide a fast and roughly sufficient estimation for PV band area. For a mask *M*, PV band area, denoted by *PV(M)*, is the XOR area between $G_i(M)$ and $G_o(M)$. The less the PV band area, the more is the mask robustness against process variations. **Figure 6** illustrates PV band area for a given mask [17].

Mask manufacturability evaluation: Mask manufacturability is evaluated in terms of satisfying mask notch and spacing design rules. The more the rule violations, the lower is the manufacturability of the mask. **Figure 7(a)** illustrates examples of design rule violations. Mask notch rule defines the minimum allowable edge length in the mask polygons, denoted by d_n. Thus, the number of mask notch rule violations of mask *M*, denoted by *#NotchV(M)*, is formulated as in Eq. (3):

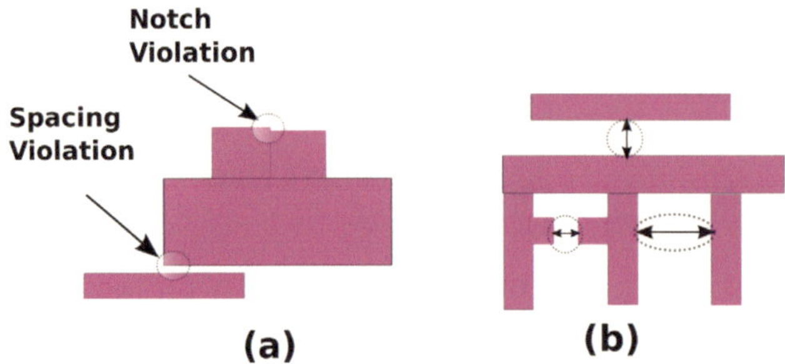

Figure 7. (a) Design rule violations and (b) comparison pair examples [16].

$$\#NotchV(M) = |\{e|e \in M; l(e) < d_n\}| \quad (3)$$

Two edges in the mask boundary violate mask spacing rule if the Manhattan distance between them is less the minimum allowable spacing distance d_s. However, both edges should either belong to two different polygons or belong to the same polygon without overlapping between them, as illustrated in **Figure 7(b)**. Such edges are said to be a comparison pair. Consequently, the number of spacing rule violations, denoted by *#SpaceV(M)*, is given in Eq. (4), where C_p represents the set of comparison pairs in M. Comparison pairs of a mask can be retrieved by bounding techniques [16].

$$\#SpaceV(M) = |(e_i, e_j)|(e_i, e_j) \in C_p, D(e_i, e_j) < d_s| \quad (4)$$

4. Tap point intensity estimation

The purpose of the proposed intensity modeling in [17] is to roughly estimate the intensity map of a mask using SOCS model within a short time. As lower weight kernel contribution in intensity value is typically small [32], such contribution in intensity value for each pixel does not dramatically change much if a mask pattern is slightly modified. On the other hand, top weight kernel contribution is significantly affected by such mask modifications. Thus, by utilizing lower weight kernel intensity information of some reference mask, the intensity map of a general mask can be estimated using only top weight kernel, followed by proper compensation with exploiting the intensity information of the reference mask.

4.1. Top weight kernel intensity modeling

Let $F_1(d)$ and $F_2(d)$ be the functions that represent the intensity impact induced by a segment to its own tap point and to the neighbor tap point, respectively, where d represents the shifting distance of that segment from its original position in the target T. The differences of intensity impact to a segment tap point and to neighbor tap point between cases when the shifting distances of that segment are d and d' are represented by $F_1(d, d')$ and $F_2(d, d')$, respectively. Let $B(w)$ represent the intensity impact induced by a serif feature on a corner to tap point t located on a corresponding corner segment, where w represents the width of the serif. The differences of intensity impact between cases when the widths of the serif are w and w' are represented by $B(w, w')$ [15].

With assuming the linearity of F_j as proposed in [15], $F_j(d, d') = F_j(d') - F_j(d) = \alpha_j$ (d' − d), where α_j is a constant ($j = 1, 2$) obtained by regression. Additionally, it is assumed that B is a quadratic function such that $B(w, w') = B(w') - B(w) = \beta (w' - w)^2 + \gamma (w' - w)$, where β and γ are constants obtained through regression.

Let $(s_0, s_1, \ldots, s_m)$ be a sequence of segments defined along the edge between corner c_0 and c_1 on the boundary of a polygon in T by fragmentation, and t_i be the tap point of s_i $(0 \geq i \geq m)$. Let d'_i and d_i be the shifting distances from the boundary in the target T for segment s_i in masks M and M_{ref}, respectively. In addition, let w_j' and w_j be the widths of serif feature on a corner c_j in masks M and M_{ref}, respectively $(0 \geq i \geq m, 0 \geq j \geq 1)$. **Figure 8** depicts the given situation. With exploiting

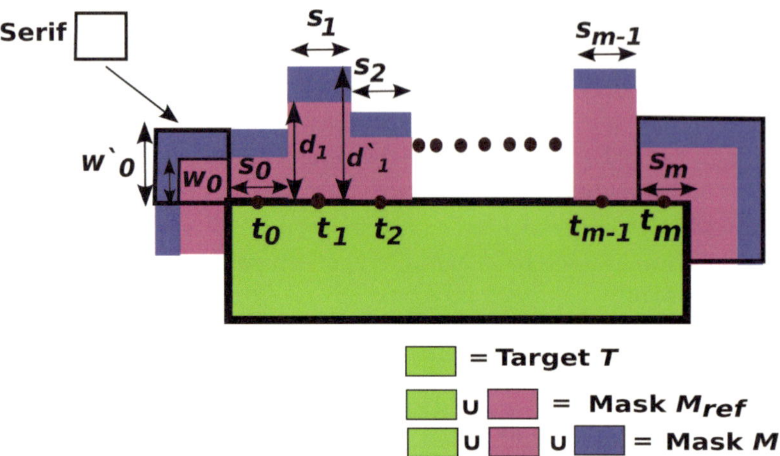

Figure 8. Top weight kernel modeling situation [17].

top kernel modeling, the intensity $I_{Pc}(M, t_i)$ of tap point t_i under process condition P_c is given in Eq. (5) for corner segment and in Eq. (6) for non-corner segment, where $\Delta x = x' - x$ [17].

$$\begin{aligned} I_{P_c}(M, t_0) &\approx I_{P_c}(M_{\text{ref}}, t_0) + B(w_0, w\prime_0) + F_1(d_0, d_0') + F_2(d_1, d_1') \\ &\approx I_{P_c}(M_{\text{ref}}, t_0) + B(w_0, w_0') + \alpha_1 \triangle d_0 + \alpha_2 \triangle d_1 \end{aligned} \tag{5}$$

$$\begin{aligned} I_{P_c}(M, t_i) &\approx I_{P_c}(M_{\text{ref}}, t_i) + F_2(d_{i-1}, d_{i-1}') + F_1(d_i, d_i') + F_2(d_{i+1}, d_{i+1}') \\ &\approx I_{P_c}(M_{\text{ref}}, t_i) + \alpha_1 \Delta d_i + \alpha_2(\Delta d_{i-1} + \Delta d_{i+1}) \end{aligned} \tag{6}$$

4.2. Lower weight kernel intensity modeling

Intensity difference map (IDM) is introduced as the mathematical difference between two intensity maps obtained using two sets of kernels [33]. Let $I_{diff}(M, K, K')$ be the IDM between intensity maps $I(M, K)$ and $I(M, K')$, where $I(M, K)$ denotes the intensity map obtained using set of kernels K and $K' \subset K$, respectively, as formulated in Eq. (7).

$$I_{diff}(M, K, K^{`}) = I_{P_c}(M, K) - I_{P_c}(M, K^{`}) \tag{7}$$

Typically, there is a trade-off between intensity map accuracy and the number of kernels used to obtain that map. However, with relaxed EPE evaluation, a set of top weight kernels in a lithographic system can be sufficient to be used for in intensity estimation, and thus, to guide the OPC response. Let K denotes the set of all kernels and $K_{suff} \subset K$ denote the set of top weight kernels roughly sufficient for optimization. Besides, let $k_0 \in K$ denote the top weight kernel [15].

In lower weight kernel modeling, intensity map for a mask M is roughly estimated through using a reference mask M_{ref} (both M and M_{ref} have been derived from the same target) as follows: The IDM of mask M_{ref} under a certain process condition is obtained using K_{suff} and

$\{k_0\}$. To estimate the intensity map of mask M, IDM works as a compensative map to the top weight kernel intensity map as given in Eq. (8). This modeling reduces effectively the simulation time since only one convolution operation is required [15].

$$I_{P_c}(M) \approx I_{P_c}(M, \{k_0\}) + I_{diff}\left(M_{ref}, K_{suff}, \{k_0\}\right) \tag{8}$$

5. OPC engine framework

Figure 9 illustrates the general framework of the OPC engine proposed in [15, 16]. Before performing the actual OPC algorithm, a preprocessing phase, wherein, the parameters that guide OPC algorithm are found through regression. The input of the OPC algorithm is a target pattern and the output is a mask solution. This algorithm consists of initialization phase, input intensity modeling, mask correction phase, mask evaluation, and post-OPC phase.

5.1. Initialization phase

This phase aims to accelerate the algorithm convergence through finding an initial mask solution whose pattern is not much deviated from the final mask solution. Initialization phase includes the following:

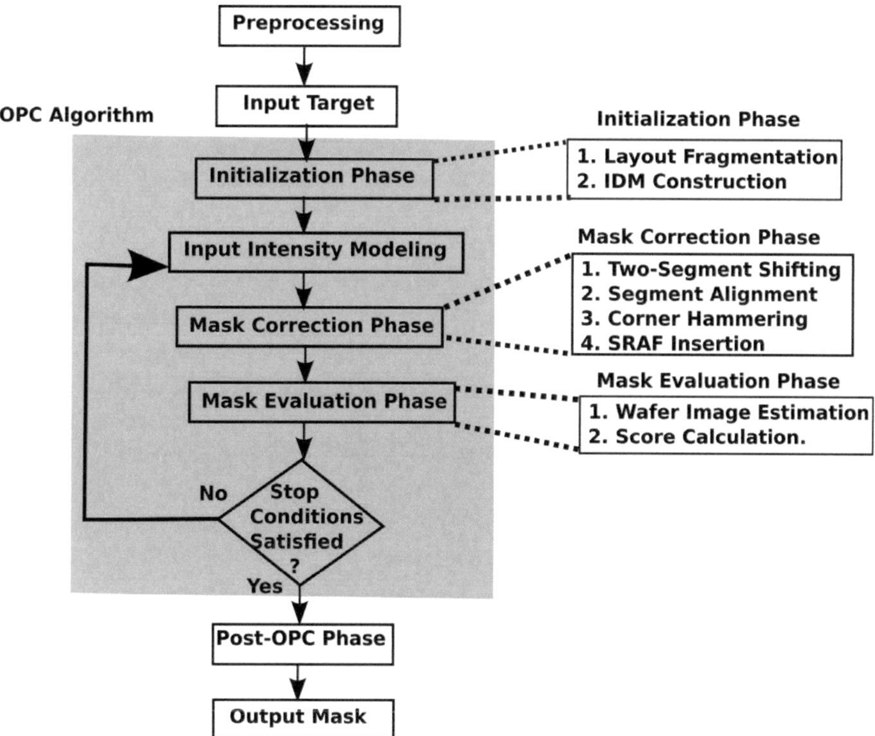

Figure 9. OPC engine framework.

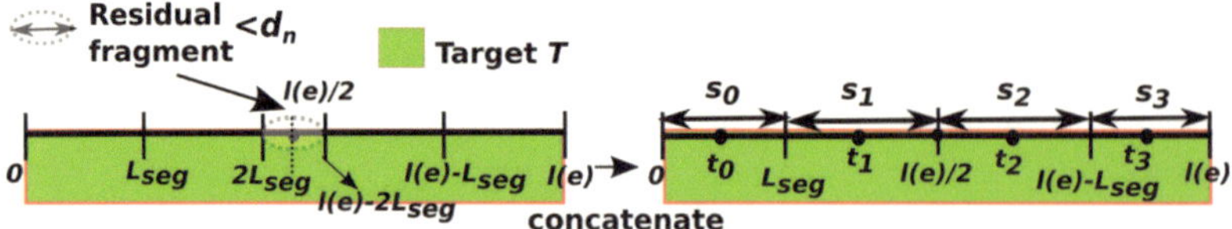

Figure 10. Fragmentation process [15].

Layout fragmentation: Edges along the boundary of target T are fragmented into segments. Segment length L_{seg} is predefined such that L_{seg} is greater than the minimum allowable notch width d_n. If a segment length is less than d_n, it is equally concatenated with its neighbors. The center for each segment s_i on the target is defined as a tap point t_i, as illustrated in **Figure 10** [15].

Intensity Difference Map (IDM) construction: One extra mask correction step is applied to generate a mask $M^{[0]}$ whose features are printable around target boundaries. With setting $M_{ref} = M^{[0]}$ and $K = K_{suff}$, IDM is constructed and exploited as in Eq. (9) to estimate intensity map of a mask M, where k_0 represents the top weight kernel in K [15].

$$\begin{aligned} I_{\text{diff}}(M^{[0]}, K_{\text{suff}}, \{k_0\}) &= I(M^{[0]}, K_{\text{suff}}) - I(M^{[0]}, \{k_0\}) \\ I_{P_c}(M) &\approx I_{P_c}(M, \{k_0\}) + I_{\text{diff}}(M^{[0]}, K_{\text{suff}}, \{k_0\}) \end{aligned} \tag{9}$$

5.2. Input intensity modeling

An OPC algorithm typically tries to make the nominal intensity curve of a given tap point crossing the target boundary at I_{th}, as depicted in **Figure 11(a)**. The distance from the target boundary to the cross-point of innermost intensity at which $I_i = I_{th}$ contributes to PV band as

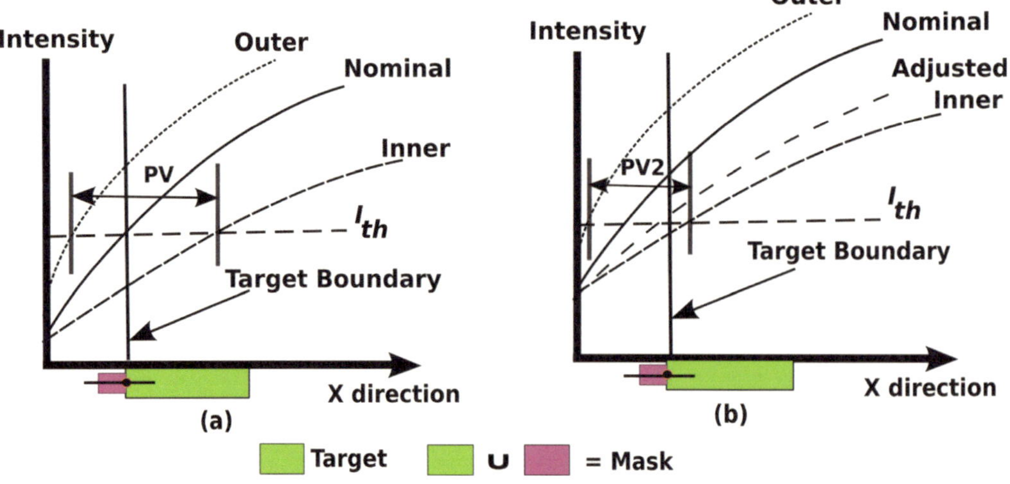

Figure 11. (a) Nominal intensity is considered to reach I_{th}, resulting in PV as PV band area indicator. (b) Adjusted intensity is considered to reach I_{th}, resulting in $PV2 < P\,V$ as PV band reduction [17].

well as distance from target boundary to the cross-point of outermost intensity. However, the innermost intensity cross-point to I_{th} is typically larger from the target boundary than the outermost intensity cross-point.

With making the cross-point of nominal intensity with I_{th} slightly outside the target boundary, PV band area can be reduced (as shown in **Figure 11b**). This is reasonable because the innermost intensity cross-point to I_{th} reaches close to the target boundary, which results in lesser PV band, since outermost intensity has already been saturated and its cross-point distance from target boundary is not expected to change significantly.

As an implementation, let $I_{def}(M)$ denote the intensity map under nominal dose and defocus. $I_n(M)$ denotes the nominal intensity under nominal dose and best-focus. In [15], the adjusted intensity map is defined, denoted by $I_{adj}(M)$, as the intensity map obtained by averaging both $I_n(M)$ and $I_{def}(M)$, as given in Eq. (10).

$$\forall p \in R, I_{adj}(M,p) = \frac{I_n(M,p) + I_{def}(M,p)}{2} \tag{10}$$

5.3. Mask correction phase

Mask correction phase applies a set of OPC steps on the input mask to optimize both EPE and PV band area with satisfying design rules. Adjusted intensity map of the input mask drives segment shifting and corner hammering, while innermost and outermost maps control SRAFs insertion.

Two-segment shifting: Let s_i and s_{i+1} be two neighboring segments with positions P_i and P_{i+1}, respectively, in mask M (see **Figure 12(a)**). The purpose is to find the new positions of those segments, denoted by P'_i and P'_{i+1}, such that the estimated intensities of their tap points become I_{th}. With exploiting top weight kernel model, the objective is to find (ΔP_i, ΔP_{i+1}) in Eq. (11) such that $\Delta P_i = P'_i - P_i$, $\Delta P_{i+1} = P'_{i+1} - P_{i+1}$. With solving Eq. (11), the new positions P'_i and P'_{i+1} are given in Eq. (12). **Figure 12(b)** illustrates two-segment shifting subroutine [15].

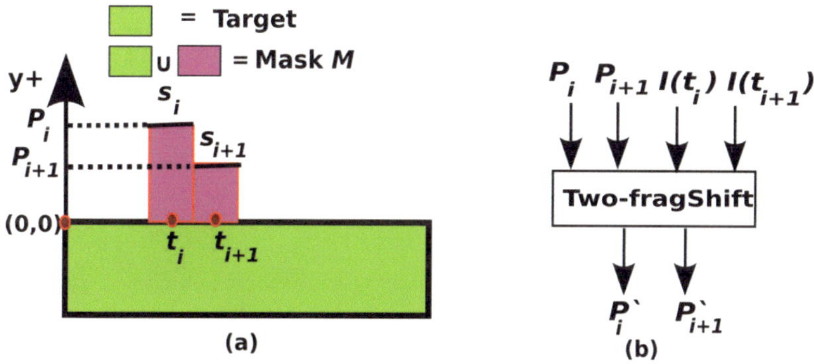

Figure 12. Two-fragment shifting: (a) current situation and (b) subroutine.

$$\begin{bmatrix} I(t_i) + \alpha_1 \triangle P_i + \alpha_2 \triangle P_{i+1} \\ I(t_{i+1}) + \alpha_1 \triangle P_{i+1} + \alpha_2 \triangle P_i \end{bmatrix} = \begin{bmatrix} I_{\text{th}} \\ I_{\text{th}} \end{bmatrix} \tag{11}$$

$$\begin{aligned} P_i' &= P_i + \frac{\alpha_1(I_{\text{th}} - I(t_i)) - \alpha_2(I_{\text{th}} - I(t_{i+1}))}{\alpha_1^2 - \alpha_2^2} \\ P_{i+1}' &= P_{i+1} + \frac{\alpha_1(I_{\text{th}} - I(t_{i+1})) - \alpha_2(I_{\text{th}} - I(t_i))}{\alpha_1^2 - \alpha_2^2} \end{aligned} \tag{12}$$

Consider the situation of non-corner segments s_{i-1}, s_i, s_{i+1} in mask M shown in **Figure 13(a)**. As illustrated in **Figure 13(b)**, Two-fragShift subroutine is applied first to s_{i-1} and s_i, followed by setting their tap point intensities to I_{th}. $I(t_{i+1})$ change due to s_i shifting is linearly estimated according to top weight kernel modeling. These data are inputted to Two-fragShift subroutine, which is then applied to s_i and s_{i+1} [15].

Corner hammering: Let c be a corner wherein corner segments a_c and b_c meet (in target T). A hammer is formed on c by shifting both a_c and b_c outside the polygon with distance w_c. This shifting amount is equivalent to the serif width as depicted in **Figure 14(a)**. Thus, the purpose

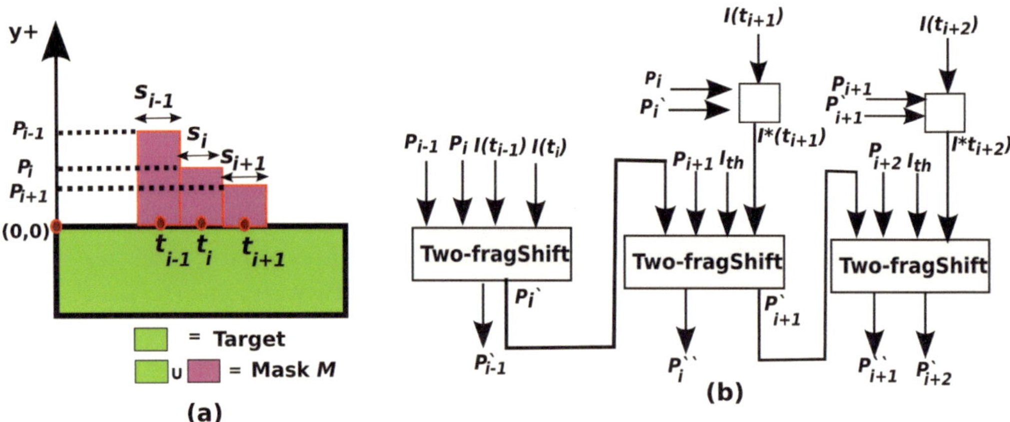

Figure 13. Edge non-corner fragments shifting: (a) situation and (b) subroutine [15].

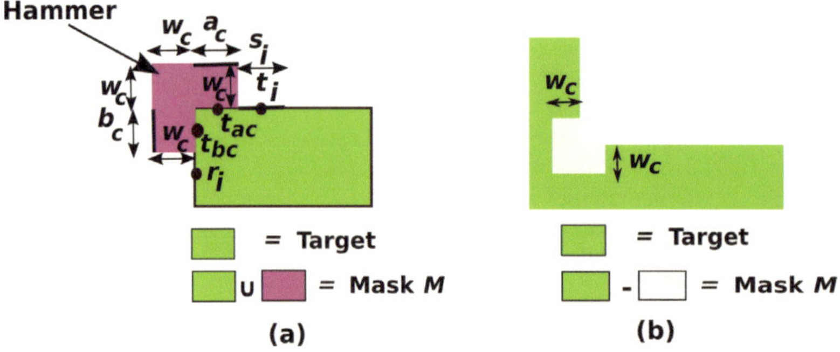

Figure 14. (a) Hammer insertion and (b) negative corner hammering [17].

is to find the serif width w'_c such that the average intensity of both corner segments tap points, denoted by t_{ac} and t_{bc}, becomes equivalent to I_{th} [15]. For a negative corner, both corner segments are shifted inside and a squared serif is picked from T, as shown in **Figure 14(b)**.

However, due to the nonlinearity of the hammering problem, several solutions might exist. However, w'_c is chosen within the interval $[w_{min}, w_{max}]$, which represents the minimum and maximum allowable serif width, where $w_{min} \geq d_n$ to satisfy notch rule and w_{max} is predefined to neglect oversized serif solutions. This problem is formulated in Eq. (13) [15].

$$\textbf{Solve for } w^*\textbf{:}\ (I(t_{ac}) + I(t_{bc}))/2 + (B(w^*) - B(w_c)) + \alpha_1 w^* = I_{\text{th}}$$
$$w'_c = \begin{cases} w^*; & w_{\min} \leq w^* \leq w_{\max} \\ w_{\max}; & w^* > w_{\max} \\ 0; & \text{Otherwise} \end{cases} \quad (13)$$

Segment alignment: Alignment aims to ensure satisfying notch rule during segment shifting. Thus, a number of parallel lines to each edge in the target are created with d_n spacing between each two consecutive lines. In this way, each segment is aligned to the closest line parallel to it after shifting, as shown in **Figure 15** [16].

SRAF insertion: With increasing the distance between an SRAF and a tap point t, the difference between outermost intensity and innermost intensity of t does not monotonically decrease. Therefore, global minimal values of this difference within the decaying intervals are SRAF candidate locations to ensure reducing $I_o(M, t)$ - $I_i(M; \text{t})$, which turns out into lesser PV band area. SRAF candidate locations are determined during preprocessing stage [15].

5.4. Post-OPC phase

Post-OPC phase aims to improve mask manufacturability through reducing mask data volume and spacing rule violations resolution. This phase consists of the following:

Segment concatenation: Reducing the segment numbers along the mask boundary helps in reducing mask data volume along with reducing the shot-count. This is achieved through two-segment concatenation. However, ad hoc concatenation of neighboring segments badly impacts pattern fidelity.

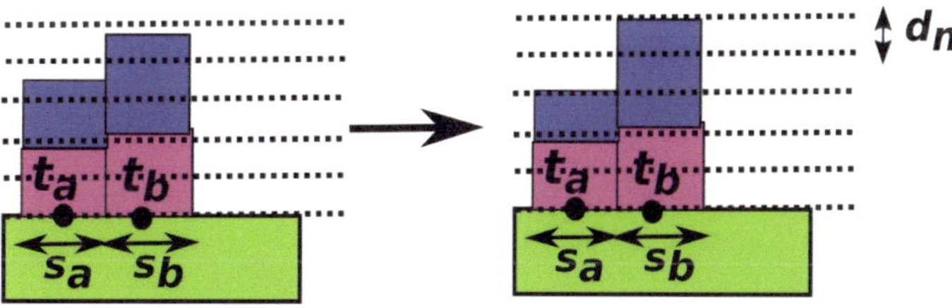

Figure 15. Segment alignment [16].

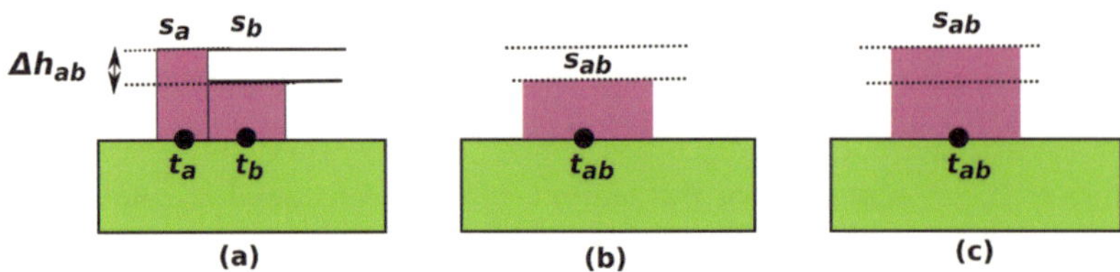

Figure 16. Concatenation process [16]: (a) before concatenation, (b) s_a is moving, and (c) s_b is moving.

Let s_a and s_b be two neighboring segments with Δh_{ab} orthogonal distance between them (**Figure 16(a)**). Let epe_{prea} and epe_{preb} denote the predicted EPE in tap points t_a and t_b, respectively, after concatenation. Concatenation process is performed as follows [16]:

- If $epe_{prea} < epe_{preb}$ and $epe_{prea} \geq epe_{max}$, shift s_a to concatenate with s_b (**Figure 16(b)**).
- if $epe_{preb} < epe_{prea}$ and $epe_{preb} \geq epe_{max}$, shift s_b to concatenate with s_a (**Figure 16(c)**).
- If the predicted EPE causes violation, no concatenation is performed.
- If concatenation is done, s_a and s_b become one segment s_{ab}.

Feature movement: Segment/SRAF extra movement aims to resolve spacing violations in the mask pattern outputted from concatenation process. This is strictly subjected to the constraint that no additional EPE violations occur, as illustrated in **Figure 17** [16].

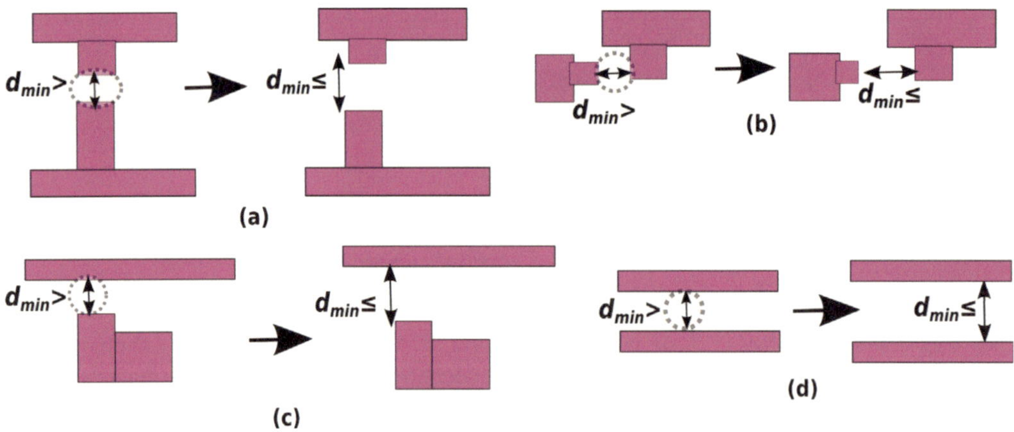

Figure 17. Spacing violation resolution cases [16]: (a) two parallel features, (b) two orthogonal features, (c) SRAF and segment, and (d) two SRAFs.

6. Experimental results and discussion

6.1. Experimental setup

Simulation environment: Lithosim uses industrial optical models with 193 nm immersion lithography. CTR model is used with intensity threshold of 0:225. Layout patterns are defined in 1024 × 1024 pixels region, where each pixel represents 1 nm × 1 nm. A set of 24 SOCS kernels forms the optical model in Lithosim [11].

OPC algorithm parameters: The proposed OPC algorithm in [16] has been implemented on top of Lithosim. The algorithm was executed on 4 cores 3.6 GHz Linux machine with total memory of 1,986,912 kB. Segment length has been chosen as 20 nm; minimum allowable mask notch has been set practically to 5 nm. The maximum allowable hammer width is 80 nm; maximum allowable SRAF width is 60 nm. The maximum number of iterations has been set to 10.

Testing benchmarks: Testing benchmarks have been provided by IBM for ICCAD 2013 CAD contest. Each benchmark is an M1 layout pattern for 32 nm technology nodes. The CD of those benchmarks ranges from 20 to 80 nm. The number of patterns (polygons) in those benchmarks ranges between 4 and 34 polygons with layout density ranges from 0.3 to 0.46 due to the pitch spacing design rules for realistic industrial cases [11].

Mask evaluation: The score function used in ICCAD 2013 CAD contest is used for evaluation [37]. Given a mask M, the score of M, denoted by $\varphi(M)$, is given in Eq. (14), where τ denotes the computation time to find a mask and ζ represents the number of hole shapes in the corrected mask. α, β, and γ are set to 5000, 4, and 10,000 following the contest.

$$\varphi(M) = \alpha^* \#EPEV(M) + \beta^* PV(M) + \gamma^* \zeta(M) + \tau \tag{14}$$

6.2. Comparison with recent algorithms

The proposed algorithm in [16] has been compared with recently published algorithms executed on the same benchmarks. **Table 1** shows a comparison between the proposed algorithm and state-of-the-art algorithms including: MOSAIC fast [36], MOSAIC exact [36], and PV-OPC [11].

The proposed algorithm in [16] outperforms MOSAIC fast in the overall score and it is 3.76 times faster. MOSAIC fast is effective in terms of PV band area due to its pixel-based behavior in finding the mask solution under each process condition. However, it has lack of estimation accuracy, which turns out into pattern fidelity degradation. MOSAIC exact effectively optimizes both EPE and PV band area since it simulates wafer image under each process condition using all kernels. However, this algorithm slowly converges. While the proposed algorithm in [16] has almost the same cost of MOSAIC exact in terms of EPE and PV band area, it is 22 times faster. PV-OPC is an effective algorithm as it exploits variational EPE under representative process conditions with satisfying mask notch rule. Keep out zone (KOZ) concept is exploited as well to avoid pinching and bridging errors between patterns. Thus, PV-OPC algorithm outperforms [16] in terms of EPE while [16] has less PV band area due to input intensity modeling and SRAFs insertion. Additionally, the proposed algorithm in [16] is 1.65 times faster. Note that PV-OPC does not consider spacing rule violations and mask data volume reduction.

Generally, it seems obvious that the proposed algorithm in [16] outperforms other recent algorithms, specifically in OPC runtime as it is 1.65 times faster than the fastest algorithm among others. Exploiting intensity difference map concept is the main reason, which turns out into minimizing the number of kernels needed for simulation during optimization.

Benchmark	MOSAIC fast				MOSAIC exact				PV-OPC				Proposed algorithm [16]			
	#EPEV	PV	Time	Score	#EPEV	PV	Time	Score	#EPEV	PV	Time	Score	#EPEV	PV	Time	Score
B1	6	58,232	318	263,246	9	56,890	1707	274,267	2	58,269	164	243,240	6	61,474	78	275,974
B2	10	47,139	256	238,812	4	48,312	1245	214,493	0	52,674	130	210,826	5	48,925	84	220,784
B3	59	82,195	321	624,101	52	84,608	2522	600,954	47	81,541	203	561,367	44	98,257	81	613,109
B4	1	28,244	322	118,298	3	24,723	1269	115,161	0	26,960	190	108,105	2	26,853	80	117,492
B5	6	56,253	315	255,327	2	56,299	2167	237,363	4	61,820	62	267,342	0	61,810	79	247,319
B6	1	50,981	314	209,238	1	49,285	2084	204,224	0	55,090	54	220,414	1	50,227	82	205,990
B7	0	46,309	239	185,475	0	46,280	1641	186,761	0	51,977	74	207,982	0	42,547	80	170,268
B8	2	22,482	258	100,186	2	22,342	663	100,031	0	22,869	65	91,541	0	22,078	69	88,381
B9	6	65,331	322	291,646	3	62,529	3022	268,138	0	70,713	55	282,907	0	65,047	75	260,263
B10	0	18,868	231	75,703	0	18,141	712	73,276	0	17,846	41	71,425	0	17,328	62	69,374
Ratio	1.57	0.96	3.76	1.04	1.31	0.95	22.12	1.0	0.91	1.01	1.65	1.0	1.0	1.0	1.0	1.0

Table 1. Comparison with state-of-the-art.

Benchmark	Algorithm published in [35]							Proposed algorithm in [16]						
	#EPEV	PV	Time	Score	#NotchV	#SpaceV	Volume	#EPEV	PV	Time	Score	#NotchV	#SpaceV	Volume
B1	0	66,218	278	265,150	186	67	10,695	6	61,474	78	275,974	0	9	5863
B2	0	53,434	142	213,878	175	68	9139	5	48,925	84	220,784	0	8	4739
B3	18	146,776	152	677,256	215	83	12,013	44	98,257	81	613,109	0	15	6902
B4	0	33,266	307	133,371	77	84	7096	2	26,853	80	117,492	0	4	2328
B5	1	65,631	189	267,713	214	124	13,582	0	61,810	79	247,319	0	8	5356
B6	0	62,068	353	248,625	224	127	13,692	1	50,227	82	205,990	0	11	5592
B7	0	51,069	219	204,495	120	134	13,019	0	42,547	80	170,268	0	3	3172
B8	0	25,898	99	103,691	109	68	7285	0	22,078	69	88,381	0	4	3027
B9	1	75,387	119	306,667	227	132	15,426	0	65,047	75	260,263	0	13	6047
B10	0	18,141	61	72,625	78	31	4934	0	17,328	62	69,374	0	0	2246
Ratio	0.36	1.21	2.50	1.10		12.24	2.36	1.0	1.0	1.0	1.0	1.0	1.0	1.0

Table 2. Mask manufacturability comparison with state-of-the-art algorithm.

To verify the effectiveness of the proposed OPC algorithm from mask manufacturability perspective, the algorithm published in [35] and the proposed algorithm in [16] have been compared in terms of mask notch and spacing rule violations, in addition to the mask data volume. **Table 2** shows this comparison, in which pattern fidelity, process variability, and computation time are included.

As shown **Table 2**, the algorithm in [35] effectively tackles pattern fidelity under nominal process condition. However, it has a relatively large PV band area. Algorithm in [16] outperforms the overall score of the algorithm published in [35] by 9%. Additionally, it is 2.5 times faster. Mask notch violations have been totally eliminated due to alignment stage while spacing violations have been reduced by 92% on average due to features movement. Mask data volume has been reduced by around 57.6% on average due to segments concatenation and alignment.

Figure 18 illustrates a target pattern, its generated mask solution using the proposed algorithm in [16], nominal wafer image, and PV band.

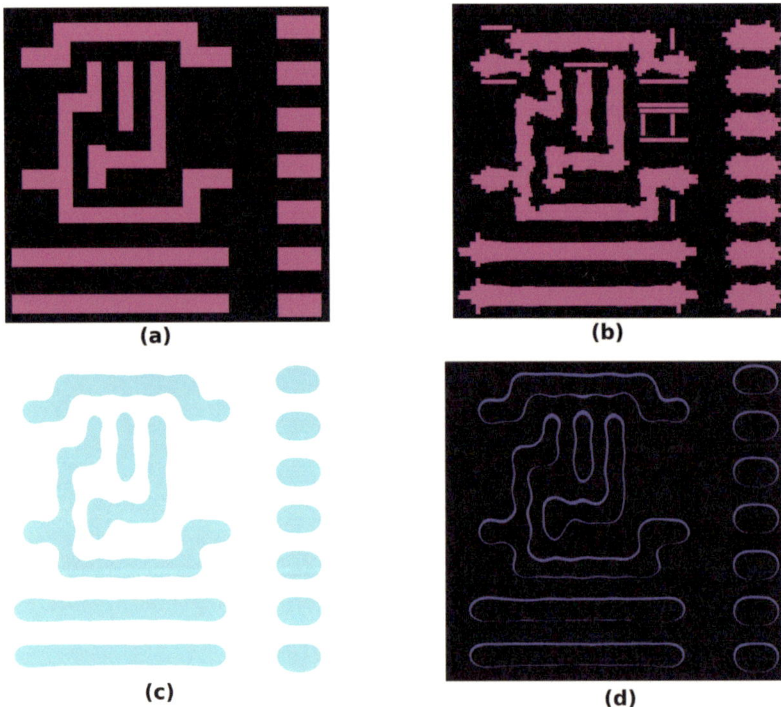

Figure 18. (a) Target pattern, (b) mask solution using [18], (c) nominal wafer image, and (d) PV band.

7. Conclusions

In this chapter, we have discussed the recent state-of-the-art OPC algorithms to tackle mask optimization problem for advanced technology nodes patterning through optical system.

Then, we have analyzed the algorithm published in [17, 18] as fast, recent OPC methodology to generate mask solutions. The analyzed algorithm outperforms other state-of-the-art algorithms in terms of EPE and PV band area reduction due to OPC adjustments guided by adjusted intensity in addition to SRAFs insertion/sizing. Computation time reduction is evident due to the fast novel intensity estimation model exploited in the OPC engine. Mask manufacturability has been significantly improved due to the post-OPC stages, wherein EPE prediction models are exploited to preserve acceptable pattern fidelity and robustness against process variations while respecting mask design rule constraints.

Author details

Ahmed Awad[1]*, Atsushi Takahashi[2] and Chikaaki Kodaman[3]

*Address all correspondence to: ahmedawad@najah.edu

1 Department of Information and Computer Science, Faculty of Engineering and Information Technology, An-Najah National University, Nablus, Palestine

2 Department of Information and Communication Engineering, School of Engineering, Tokyo Institute of Technology, Tokyo, Japan

3 Toshiba Memory Corporation, Kanagawa, Japan

References

[1] Xu M, Arce G. Computational Lithography. Wiley Publisher; 2010

[2] Mack C, Carback R. Modeling the effects of prebake on positive resist processing. Proceedings of Kodak Microelectronics Seminar Interface. 1985. pp. 155-158

[3] International Technology Roadmap for Semiconductors. Technical Report. 2014. http://public.itrs.net

[4] Wong B, Mittal A, Starr G, Zach F, Moroz V, Kahng A. Nano-CMOS Design for Manufacturability: Robust Circuit and Physical Design for sub-65nm Technology Nodes. Wiley Publisher; 2008

[5] Mack C. Corner rounding and line-end shortening in optical lithography. Proceedings of SPIE. 2000;**4226**:83-92

[6] Harriott L. Limits of lithography. Proceedings of IEEE. 2002:366-374

[7] Wei Y, Back D. 193 nm Immersion Lithography: Status and Challenges. SPIE Newsroom; 2007

[8] Shibuya M. Resolution Enhancement Techniques for Optical Lithography and Optical Imaging Theory. Optical Review. 1997

[9] Su Y-H, Huang Y-C, Tsai L-C, Chang Y-W, Banerjee S. Fast lithographic mask optimization considering process variation. IEEE Transactions on Computer Aided Design of Integrated Circuits and Systems. 2016

[10] Awad A, Takahashi A, Tanaka S, Kodama C. A fast process variation and pattern fidelity aware mask optimization algorithm. Proceedings of ICCAD. 2014. pp. 238-245

[11] Banerjee S, Li Z, Nassif S. CAD contest in mask optimization and benchmark suite. Proceedings of ICCAD. 2013. pp. 271-274

[12] Word J, Mizuuchi K, Fu S, Brown W, Sahouria E. Mask shot count reduction strategies in the OPC flow. Proceedings of SPIE 7028, Photomask and Next-Generation Lithography Mask Technology XV. 2008

[13] Tanaka S, Inoue S, Kotani T, Izuha K, Mori I. Impact of OPC aggressiveness on mask manufacturability. Proceedings of SPIE 5130, Photomask and Next-Generation Lithography Mask Technology. 2003

[14] Cobb N, Zakhor A. Fast sparse aerial image calculation for OPC. Proceedings of SPIE. 1995;**2621**:534-545

[15] Awad A, Takahashi A, Tanaka S, Kodama C. A fast process variation aware mask optimization algorithm with a novel intensity modeling. IEEE Transactions on Very Large Scale Integration Systems (TVLSI). 2017;**25**(3):998-1011

[16] Awad A, Takahashi A, Kodama C. A fast mask manufacturability and process variation aware OPC algorithm with exploiting a novel intensity estimation model. IEICE Trans. Fundamentals. 2016;**E99-A**(12):2363-2374

[17] Cobb N, Granik Y. Model-based OPC using the MEEF matrix. Proceedings of Annual BACUS Symposium on Photomask Technology. 2002. pp. 1281-1292

[18] Lei J, Hong L, Lippincott G, Word J. Model-based OPC using MEEF matrix II. Proceedings of SPIE 9052, Optical Microlithography XXVII. 2014

[19] Fuhner T, Erdmann A. Improved mask and source representations for automatic optimization of lithographic process conditions using a genetic algorithm. Proceedings of SPIE 5754, Optical Microlithography. 2005

[20] Yu P, Pan D. A novel intensity based optical proximity correction algorithm with speedup in lithography simulation. Proceedings of ICCAD. 2007. pp. 854-858

[21] Mukhejree M, Buamm Z, Lavin M, Samuels D, Singh R. Method for adaptive segment refinement optical proximity correction. US Patent 7043712. 2006

[22] Yu P, Shi S, Pan D. Process variation aware optical proximity correction with variational lithography modeling. Proceedings of DAC. 2006. pp. 785-790

[23] Krasnoperova A, Culp J, Graur I, Manseld S, Al-Imam M, Maaty H. Process window OPC for reduced process variability and enhanced yield. Proceedings of SPIE 6154, Optical Microlithography XIX. 2006

[24] Agarwal K, Banerjee S. Design driven patterning optimization for low K1 lithography. Proceedings of IEEE International Conference on IC Design & Technology. 2012. pp. 1-4

[25] Banerjee S, Agarwal K, Orshansky M. SMATO: Simultaneous mask and target optimization for improving lithographic process window. Proceedings of ICCAD. 2010. pp. 100-106

[26] Pang L, Liu Y, Abrams D. Inverse lithography technology (ILT), what is the impact to photomask industry?. Proceedings of SPIE 6283, Photomask and Next-Generation Lithography Mask Technology. 2006

[27] Liu Y, Abrams D, Pang L, Moore A. TIP-OPC: Inverse lithography technology principles in practice: unintuitive patterns. Proceedings of SPIE. 2005;**5992**:886-893

[28] Mulkherjee M, Manseld S, Leibmann L, Lvov A, Pa-padopoulou E, Lavin M, Zhao Z. The problem of optimal placement of sub-resolution assist features (SRAF). Proceedings of SPIE 5754, Optical Microlithography, 2005

[29] Manseld S, Liebmann L, Molless A, Wong A. Lithographic comparison of assist features design strategies. Proceedings of SPIE. 2000;**4346**:63-76

[30] Capodieci L, Gupta P, Kahng A, Sylvester D, Yang J. Toward a methodology for manufacturability-driven design rule exploration. Proceedings of DAC. 2004. pp. 311-316

[31] Gupta P, Kahng A, Muddu S, Nakagawa S, Park C-H. Modeling OPC complexity for design for manufacturability. Proceedings of SPIE. 2005;**5992**:1-9

[32] Gallatin G, Lai K, Mukhejree M, Rosenbluth A. Printability verification by progressive modeling accuracy. US Patent 7512927. 2009

[33] Awad A, Takahashi A, Tanaka S, Kodama C. Intensity difference map (IDM) accuracy analysis for OPC efficiency verification and further enhancement. IPSJ Transactions on System LSI Design Methodology. 2017;**10**:28-38

[34] Gao J-R, Xu X, Yu B, Pan D. MOSAIC: Mask optimization solution with process window aware inverse correction. Proceedins of DAC. 2014. pp. 1-6

[35] Kuang J, Chow W-K, Young E. A robust approach for process variation aware mask optimization. Proceedings of DATE. 2015. pp. 1591-1594

[36] Acosta C, Salazar D, Morales D. A novel algorithm for notch detection. Proceedings of SPIE. 2013;**8701**

[37] http://cad_contest.cs.nctu.edu.tw/CAD-contest-at-ICCAD2013/problem_c/

[38] Cobb N. Sum of Coherent Systems Decomposition by SVD, Berkeley CA. 1995. pp. 1-7